Managing Business in a Multi-Channel World:

Success Factors for E-Business

Timo Saarinen
Helsinki School of Economics, Finland

Markku Tinnilä
Helsinki School of Economics, Finland

Anne Tseng
Helsinki School of Economics, Finland

IDEA GROUP PUBLISHING
Hershey • London • Melbourne • Singapore

Acquisitions Editor:	Renée Davies
Development Editor:	Kristin Roth
Senior Managing Editor:	Amanda Appicello
Managing Editor:	Jennifer Neidig
Copy Editor:	Sue VanderHook
Typesetter:	Cindy Consonery
Cover Design:	Lisa Tosheff
Printed at:	Yurchak Printing Inc.

Published in the United States of America by
Idea Group Publishing (an imprint of Idea Group Inc.)
701 E. Chocolate Avenue
Hershey PA 17033
Tel: 717-533-8845
Fax: 717-533-8661
E-mail: cust@idea-group.com
Web site: http://www.idea-group.com

and in the United Kingdom by
Idea Group Publishing (an imprint of Idea Group Inc.)
3 Henrietta Street
Covent Garden
London WC2E 8LU
Tel: 44 20 7240 0856
Fax: 44 20 7379 3313
Web site: http://www.eurospan.co.uk

Library of Congress Cataloging-in-Publication Data

Managing business in a multi-channel world : success factors for E-business / Timo Saarinen, Markku Tinnilä and Anne Tseng, editor.
p. cm.
Summary: "This book is designed to address the concerns of existing companies who wish to succeed in the new multi-channel environment as it develops and becomes commonplace"--Provided by publisher.
Includes bibliographical references and index.
ISBN 1-59140-629-3 (hard cover) -- ISBN 1-59140-630-7 (soft cover) -- ISBN 1-59140-631-5 (ebook)
1. Electronic commerce--Management. 2. Technological innovations--Management. 3. Information technology--Management. I. Saarinen, Timo. II. Tinnilä, Markku. III. Tseng, Anne.
HF5548.32.M354 2005
658.8'72--dc22
2005004541

British Cataloguing in Publication Data

A Cataloguing in Publication record for this book is available from the British Library.
All work contributed to this book is new, previously-unpublished material. The views expressed in this book are those of the authors, but not necessarily of the publisher.

Managing Business in a Multi-Channel World: Success Factors for E-Business

Table of Contents

Preface

Why is the Multi-Channel World Interesting?

Early predictions of the Internet and mobile services heralded a drastic change in the way we conduct business when, in fact, the opposite has occurred. Instead of replacing earlier business models, multi-channels are fast becoming an important part of existing companies' strategies and complementing the traditional channels. Rather, as the information-based society continues to develop and mature, consumers and businesses face an increasing array of choices with which to shop, purchase, obtain information, and conduct business by means of several different access devices and technologies. For these reasons, we predict that the multi-channel world will continue to grow in interest for managers and academic researchers.

The technologies and applications of multiple channels have progressed quickly and businesses have learned to make more user-friendly decisions and to adapt and integrate new and different technologies at a fast pace. Pioneering companies have learned to utilize the Internet as a channel and to develop company networks that offer integrated services or serve as intermediaries. Commercial applications for network technologies have matured and become more advanced, especially in business-to-business (B2B) commerce, where functional, integrated products or packaged services are already being used to increase sales and to improve deliveries and acquisitions. Many companies offer electronic services of their own and have increased their know-how in e-business from the necessary background systems and processes to developing appropriate, working business models. Moreover, both consumers and suppliers have grown increasingly receptive to using different channels, as access to new technologies has rapidly spread to millions of consumers. On the other hand, high development costs and low, short-term yields have driven many companies to reconsider their electronic solutions. For some, the electronic channel is seen as a value in itself and almost trendy to have, and less so because it's useful. As a result, the expected positive network effects have not been realized.

Thus, with the challenges entailed, it is our anticipation that the multi-channel world will continue to intrigue and fascinate researchers and managers. We believe that our book, *Managing Business in a Multi-Channel World,* will help managers to understand how forward-thinking organizations can best build competitive advantage from new multi-channel e-business models.

What Do We Mean by *Multi-Channel*?

While the definition of multi-channel continues to evolve with new technologies, what is clear is that what the Internet has made possible on our desktops will soon be technically possible, irrespective of the mobile device we carry or the equipment that we use at home or work. So, for the purposes of our book, we have chosen a broader definition of *multi-channel* and *multi-channel services* to mean any device or means by which companies or consumers use the Internet either to provide or access services that have traditionally been conducted manually using non-Internet equipment in person or at a shop. This may include mobile telephony and wireless data transfer technologies, such as SMS and the mobile Internet, digital or cable television, PC Internet, and even machine-to-machine transmissions.

What are the Visions of Multi-Channel E-Business, and What Kind of Development has Already Happened?

In the past, businesses have viewed new channels of electronic environment more from a technological perspective. However, in order to remain competitive and to attract and retain customers, it is becoming increasingly important for companies to broaden their perspective from a purely technical focus to more of a strategic focus. In fact, e-business is being rapidly transformed from a technology- to a business-driven mode.

Moreover, businesses face the often complex and differing task of catering to customers who may display markedly dissimilar preferences, habits, and financial backgrounds. The essential requirements for fulfilling the needs of these individual consumers are different from the requirements of business and organizational customers and represents challenges from both a technological and strategic perspective. From a technical perspective, the questions include how to seamlessly provide access to services ubiquitously (e.g., on the road, at home, at the office, in an elevator, underground, in an airplane). Moreover, from a strategic or service perspective, the challenges include understanding potential customers' interests and the conditions under which they prefer to access the multi-channel services.

In order to provide quality services in multiple channels to customers, it is apparent that a large network of organizations, both public and private, is required. This network could be easily controlled and maintained, perhaps, if the world were a constant. However, with new technologies constantly emerging, new devices being invented, and commercialized new services being developed and dropped from the marketplace, the continuously changing environment requires its participants to constantly manage and intermediate among different players in the multi-channel environment.

Development of Multi-Channel E-Businesses

Historically, the development of a company into a multi-channel e-business has involved in several phases. First, a company creates an electronic transaction, often in-house. Next, the company tries to integrate its back-office service processes with its front-end service offerings across different channels. Finally, it becomes a multi-channel e-business when it manages to successfully integrate the value chain across different channels in order to suit different customer's needs (see Figure 1).

Figure 1. Phases of e-business development

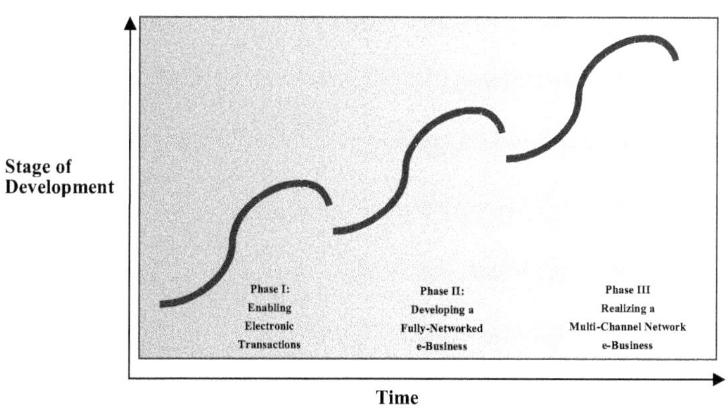

Many companies begin their e-businesses by making everything themselves, even though they may not have any previous experience in this area. This company focus has not always been the best operational model, because companies tend to be less efficient, and their electronic solutions separate from their business processes (e.g., separate ordering systems for traditional vs. electronic channels). While this may work in the short term, in the long term, electronic solutions built in this manner are unsustainable, because the majority of the electronic services answer to the needs of only a few consumers (e.g., pioneers or early adopters) and not the majority.

In the second phase of e-business development, companies often try to integrate their own internal processes and that of their service processes with electronic services. In this stage of development, companies tend to be focused on developing better integrated front-end and back-end electronic systems; however, their services are still not necessarily focused on the needs of the customers. The electronic service thus becomes an independent and functional part of a company's operations. However, as the e-business develops, new questions may start to arise as to the role of the electronic channel in the comprehensive strategy of the company and how to integrate customers across different channels.

In the final phase of development, the company moves from building individual channels (i.e., the Internet, mobile phone, etc.) to building the entire multi-channel environment. In a networked, multi-channel environment, the need for co-ordination has given rise to different mediators, whose tasks vary from information transmission to logistics.

Building a Successful Multi-Channel E-Business

As companies transition to integrated, multi-channel environments, they need to be able to build services across different channels in order to suit different customers' needs and to integrate into a network the role of every company involved in the value chain. The key features of a successful multi-channel e-business include:

- Customer focus and utility
- Maturity of technology and services
- Developing the business model
- Networking and outsourcing

In the following section we deal with these factors in more detail.

Customer Focus and Utility

When it is a question of new technologies or the services, a company must understand how the customer learns to use new devices and the customer's ability to accept and adopt a new behavior or habit. For example, it may be misleading to say that because of the high penetration rate of mobile phones, people will start using mobile phones to pay for services or products. Paying with mobile phones requires changing accustomed habits of making a purchase. For many consumers, the mobile phone is mostly a means of communication, while purchases are still made the traditional way. Yet, another example is the fact that in Europe, text message services have become popular, but other services remain unsuccessful.

When services that are based on a new technology are brought to market, one of the challenges faced by providers is matching the expectations of the service providers with that of customers' needs. Companies have launched a host of new electronic services for different customer groups (from consumers to companies), but the biggest problem has been that they lack the quality and usability needed for the mass market. One problem with cooperative networking between companies has been that every company has wanted to keep the customer relationship for itself, often resulting in inconvenience to the end-user.

Maturity of Technology and Services

When one rushes to bring a service to market, the services are not necessarily ready, even though the technology exists. This often leads to the situation in which only the pioneer customers are interested and able to use the technology and the services, whereas the mass market is likely not willing to use it. Another challenge is to develop services that can be used on multiple platforms. Consumers may not want to use mobile phone services or services of the digital television separately, but there must be a clearly logical connection of these channels to each other. Generally speaking, reaching critical mass requires that there be common standards upon which users can access the same services. Finally, the regulation of the service still may be an issue after the service is made widely available. Legislators would have to create the consumer protection regulations, within which the different parties of the trade can transact such that consumer rights are protected and the responsibilities of the companies are clearly identified and maintained.

Developing the Business Model

The search for optimal business and earnings models is the main objective of many e-businesses at the moment. When risk investments have ended, companies must find sustainable earnings models, since, in the long run, companies must be able to deliver a profit. Many different business models have been attempted but have failed because of weak foundations, especially among Internet companies. Even ideas without exact definitions of the service provided or customer segments have been called business models.

With the Internet and mobile phone services, revenue and earnings model remains a problem. In spite of an abundant service supply, the potential customers have not considered the services valuable enough to pay for them. Moreover, services often have been brought to market too early while they are still under development, and, therefore, their benefits (in practice) have been limited. Considering these points, the pricing of services plays an important role in the business model.

Networking and Outsourcing

Among multi-channel e-businesses, the focus of company operations is changing toward a more open and network-like model. When companies start to develop different services for many channels, all the services will not be worth developing and producing alone, as it has been possible via individual channels. Because one company does not have resources to produce everything itself, a coordinator of some kind will be needed to implement effective multi-channel services through network partnerships. Until now, all the big actors have wanted to be leaders of the operation in the Internet and to develop service packages by themselves. In a network model, instead, the companies' roles are not as hierarchical. However, some companies will remain in a stronger position as a central actor or subcontractor (e.g., Nokia, NTT DoCoMo).

Networking emphasizes the differentiation of companies' roles within the value chain. The value chains of the companies that operate in electronic channels can be quite new, in which case the tasks and roles of individual companies and their competitive positions within the industry sector may change considerably. Finally, as e-business develops, the roles of companies will become differentiated, and the companies likely will outsource their functions to secondary companies in different service channels. In the multi-channel environment, the need for coordination of the services of channels will give rise to a new type of service mediators, whose tasks vary from transmitting information to logistics. Specialization in the network is resulting in outsourcing of non-core activities. Many companies are outsourcing their non-core functions (e.g., logistics solutions and internal office services) and are choosing, instead, to concentrate on their core competencies.

In Table 1, we present the essential features of the development of e-business. Each feature will vary according to the company's stage of development. For example, the strategic focus of the phase of development will depend on the stage of the development. Correspondingly, the typical organization of the electronic business and its integration objectives depends on the phase of development. The organization and integration in the first stage will change when the development proceeds toward increased network cooperation and integration.

Table 1. Essential features of the phases of e-business development

	Electronic Trade	**Electronic Business**	**Multi-Channel E-Businesses**
Strategic focus	Individual solutions and services of the electronic trade. The opportunities to electronic trade are opened.	Coordination of processes inside the company services of the electronic trade. Improvement of the competitive ability of the company.	Services that are wider and more easily reached together with network partners.
Organization	E-commerce is treated as a separate project from other business strategies. Electronic commerce unit or individual business unit.	Organizing and coordination at the enterprise level	Network of the companies in the cooperation.
Integration objectives	Integration of the functions and of different systems.	Integration at the enterprise level.	Integration of the network.
Character of the innovation and challenges	Innovative solutions of the electronic trade or services to the market fast.	Innovative use of the electronic business in the core business.	New roles of companies, again the value chain's basing and coordination.
Customer relationships	Separate client connections. Emphasis on number of visitors. Collecting information through registration.	Companies stay close to the customers. How is customer relationship commanded and who maintains it?	Customer relationships are coordinated and consolidated by an infomediary or by stores through several separate channels.
Information processing systems and technologies	E-marketplace applications. Separate information processing system of the electronic trade. Other information processing systems of the company.	System integration at the company level. Integrated ERPs. Separate systems of mobile phone services. External integration of the company difficult because of missing standards and interfaces.	Several connection technologies required by the multi-channel operations. Intelligent channels, among others, the customer's identification. Common protocols, standards, and bases.
Business processes	Separate processes of the electronic trade. Separate ERP processes. Own distribution system. Drawing of functions inside company.	Integrated main processes. Integrated service processes into the background processes of companies. Intensification of the operation and removal of bottlenecks—developing of processes.	The company limits exceed the divided processes. Outsourced processes. Demands of the cooperation of different channels to the service processes. Technical redefinition of electronic processes.

Looking to the Future

In the future, cross-channel interoperability of services will be an important element of a successful supply chain and channel management. Multi-channel access likely will continue to evolve and may take the form of wearable technology or radio frequency identification (RFID) tags that will improve monitoring in the procurement process. What is clear is that understanding consumer preferences will become a critical factor upon which companies will be able to more rapidly develop new services and service features.

Themes

In this book, we will focus on three main themes: (1) the multi-channel environment from the consumer and business perspectives; (2) the markets and strategy and network of multi-channel e-business and (3) the future of the digital world. Each theme is discussed in several chapters by different authors.

In Section I, we address multi-channel usage from a consumer's perspective. Hara et al. will start by using several frameworks (e.g., BUMMAT and Future value chain) to analyze the future of mobile services. Tinnila et al. will then discuss the importance of understanding how consumers will respond to new services by surveying Finnish consumers and their preferences. Mallat and Dahlberg will assess the potential of mobile payment based on the results of empirical studies conducted on prospective consumers and merchants. Then, Öörni will examine the determinants of using Internet-based information sources in the context of tourism and travel. Finally, Elkelä et al. will discuss and compare traditional and new forms of communication (e.g., letters and e-mail), especially from the consumer point of view, and analyze their effects on the communication service evolution and consumer behavior.

From a business perspective, Merisavo will examine how marketers can use digital channels to develop and strengthen customer relationships. Holmqvist and Pessi will introduce us to the results of several years of collaborative research on implementation projects for spare parts distribution in the automotive industry. Then, Penttinen and Saarinen will describe how one specific firm has undergone the change from product-oriented thinking toward service-based concepts. Finally, in this section, Nilsson will look at the potential business value of a design-oriented research project.

In the Market, Strategy, and Network section (Section II), we will introduce original theoretical frameworks and apply them to case studies around the world. We will then discuss researchers' views on business success strategies followed by our ongoing study on the role of networks. We will first walk-through the opportunities and lessons of new markets. This will include discussion of new types of multi-channel services such as Pelkonen's discussion of mobile gaming or Orava and Perttula's description of the digital television channel, which is currently under development in Finland. However, with opportunity will likely come regulation, as Laine and Heikkilä will show by examining the regulation of online gambling from a European context.

We will next look at strategies for developing e-businesses. Klein et al. have developed a framework for multi-channel strategies that is based on channel integration and channel domination as classification criteria and on the fact that multi-channel strategies can fundamentally differ regarding the relationship of online and offline channels. Tseng et al. has compared the key characteristics of operator-driven business models in four regions around the world to develop a framework for the successful design and implementation of mobile data services, and Rajala et al. look at different channel options for distributing games for mobile devices and their related earnings logic and develop a model of channel choice.

Regarding the role of networks, Heikkilä et al. discuss the creation of a joint business model as a necessary means by which companies can coordinate co-operation among members of a collaborative network in practice. Van Heck et al. use a conceptual framework that identifies four factors that could explain the adoption of multi-channel technologies to analyze five Finnish consumer auction markets.

In our last section, we will attempt to look into the future of the digital world. We will start by examining new frameworks or intermediaries. Lähteenmäki, et al. explore the emergence of intermediaries (i.e., infomediaries) and demonstrate their development by analyzing exemplary cases. In addition, Peffers and Tuunanen introduce the Critical Success Chains approach to developing new services and use a case example to illustrate the method.

In addition, the discussion will include services of the future. Tinnilä and Lauraeus-Niinivaara will provide a vision of 2020, a consensus of the future by experts in Finland. Then Manninen and Viherä will describe the shape and development of consumers of the future through the use of storytelling. Finally, in our last chapter, Palmer concludes by describing some of the opportunities in channel management at multiple levels, including customer-focused channels providing capabilities in managing customer contacts, sales, and service, as well as procurement and distribution channels.

Acknowledgments

The thought of collecting the different ideas surrounding the multi-channel business environment of the future arose from the belief that the many interesting discussions we have had with the contributors of this book would also be of interest to others. Therefore, we would like to acknowledge the efforts of our 40 contributors for helping us make this book possible, as this has been as much their book as it is ours. We would like to recognize them for their pioneering views, which have enlightened us and from which we hope readers can find new ideas with which to solve the challenges they face in their own research or businesses.

We would also like to thank our colleagues at the Helsinki School of Economics in Finland and LTT Research, Ltd. for its support, not only during the editing of this book, but also in leading the way in developing new ideas regarding the multi-channel electronic business environment. Special thanks, in particular, go to Tommi Vihervaara for his assistance.

Furthermore, we'd like to thank our friends and business colleagues at home and abroad for their constructive comments and suggestions. Many business managers have greatly supported our efforts in understanding the phenomena of transformation in their businesses. We are greatly in their debt for bringing us down-to-earth from our theoretical spheres. Their views have increased our understanding about how new services are being developed and what kinds of ideas we can expect will materialize in the real world. We hope that they, too, have been able to benefit from our research.

We also wish to thank the staff at Idea Group Publishing and, in particular, Jan Travers and Michele Rossi, for their patience and guidance throughout the process of writing this book.

We would like to thank the Wihuri Foundation, whose funding has made the publication of this book possible.

Finally, we would like to thank our families for their love and support throughout this project: Pirkko, Olli, Sanna, and Anne; Kristina, Martin, and Henrik; and Steve.

Timo Saarinen, Markku Tinnilä, and Anne Tseng
Helsinki, Finland

Section I

Multi-Channel Perspectives

Chapter I

Introduction: Toward Seamless Multi-Channel Services

Veikko Hara, TeliaSonera AB, Sweden

Karri Mikkonen, TeliaSonera AB, Sweden

Timo Saarinen, Helsinki School of Economics, Finland

Markku Tinnilä, Helsinki School of Economics, Finland

Jarkko Vesa, Helsinki School of Economics, Finland

Abstract

The convergence of information technology, communications, and content raises the question of how service and content providers can best guarantee a seamless customer experience in a multi-service, multi-channel, and multi-device business environment of tomorrow. In this chapter, we argue that one way to achieve this goal is for service and content providers to join forces with new types of channel partners called service mediaries. In this chapter, we analyze the current structure and future trends of multi-channel service delivery by using two frameworks: the BUMMAT model (Kallio et al., 1999), which focuses on service delivery from end-user perspective; and the Future Value Chain framework (Hara, 1999), which describes the layered structure of the future service delivery business. By combining these two perspectives, we offer an in-depth analysis of the various tasks and processes that are needed in order to deliver the new kinds of services and content in a multi-channel context. We will discuss the implications of these findings for telecom operators who are interested in transforming their business models from the voice- and access-centric paradigm to a data- and service-centric business model, where a company's core competence will be in managing customer information instead of managing access networks. Our objective also is to help service and content providers to understand the current development in the telecom market, so that they have better chances to position themselves in the emerging value networks of digital services.

Introduction

For years, we have heard stories of how computing and telecommunications are becoming one and how traditional business computing and the world of consumer electronics are finally converging into something bigger, more exciting, and more rewarding than traditionally separate industries. Now, it seems that these visions are finally coming true, as we are witnessing how content and services are increasingly digital and transferred over IP-based networks, or, as the CEO of Intel Craig R. Barrett has put it, "After 20 years of talking, this so-called convergence of computing and communications is happening" (*BusinessWeek*, March 8, 2004).

Visionaries within the telecom industry anticipated the current development years ago, but, once again, the old wisdom presented by MIT professor Licklider in the 1960s appears to be true: "People tend to overestimate what can be done in one year and underestimate what can be done in five or ten years" (J.C.R. Licklider, 1965). This 10-year rule of thumb introduced by Licklider appears to apply well to the introduction of many new technologies such as fax machines, CD players, and the Internet (Odlyzko, 2001). If we take a closer look at the technological milestones behind the Internet revolution, we can see that the Internet became commercially available in 1988, followed by the World Wide Web around 1991 and easy-to-use browsers in 1993 (Dutta & Roy, 2003). It is easy to see that without these and many other inventions and innovations, the Internet would not have become such a big part of our everyday lives. Unfortunately, venture capitalists and investors often expect emerging technologies and business models to materialize much faster than what the lessons from earlier experiences would suggest, as the infamous hype curve by Gartner Group demonstrates (see Figure 1).

There are, of course, several reasons for the slower-than-expected diffusion of new technologies and services. Some of these obstacles are purely technical or

Figure 1. The e-business hype curve (Gartner Group, 1999)

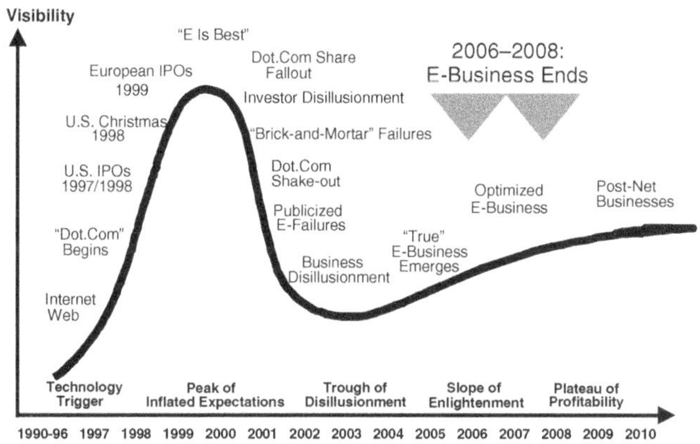

related to the increasingly complex process of creating standards that need to be widely supported and accepted in different markets. Another challenge is that new technologies and services, such as the mobile Internet or interactive digital TV, are by definition very complex products that require close cooperation among multiple players (i.e., network vendors, terminal vendors, regulators, content providers, and the developers of service delivery platforms). In order to succeed in this kind of business environment, companies need to build networks of interrelated companies that complement each other's products and services. These kinds of networks aren't created in a day, as experiences of the Japanese mobile Internet market have shown (Funk, 2004).

Let's take a closer look at the role of a telecom operator in today's business environment in a typical European market. As Figure 2 demonstrates, an operator's business model consists of multiple levels, each of which contains various kinds of tasks and processes. Starting from the top, an operator needs to work closely with its consumer and business customers in order to develop a strong demand for the services offered and in order to gain a clear perception of customers' needs and desires. The second layer focuses on the creation and management of service offering along with various kinds of value-adding functions such as packaging and bundling of online content. The mechanisms for collecting the information required to conceive an accurate customer perception are included in the processes and systems of customer automation.

The role of service mediary functions in this layered model is to provide access to various kinds of networks and to participate in the development of customer demand. Last, but not least, an operator needs to have the skills and capabilities to analyze and utilize the valuable information that flows through the business systems. We call this function the *operator brains* in order to emphasize that

Figure 2. The multiple layers of telecom operator business model

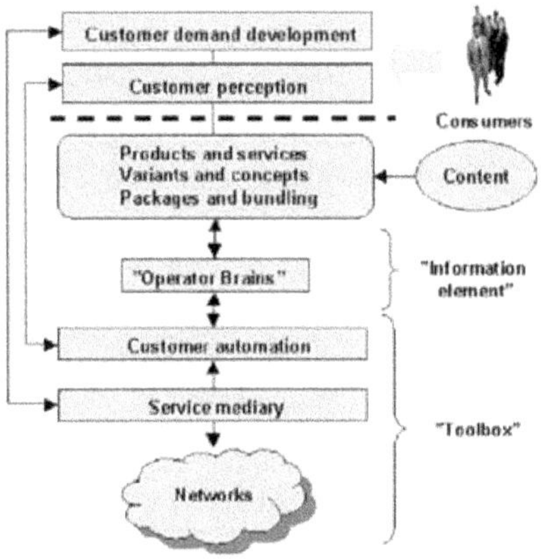

the success of this kind of business model depends on the ability of the operator to manage the information element of the model.

It is very easy to see that building the kind of business model presented in Figure 2 exceeds in many ways the scope of most traditional players' business models. What is even more exciting about the current development in the transition toward multi-channel digital content and services is that the emerging business models do not care about the traditional industry boundaries. In this chapter, we argue that this kind of development calls for totally new kinds of players who can help content and service providers move to the new world of seamless customer experience over multiple channels and devices. We call them *service mediaries*, as they represent a combination of traditional channel intermediaries (Etgar & Zusman, 1982; Rosenbloom, 1999; Stern & El-Ansary, 1998; Stigler, 1951) and infomediaries (Hagel & Rayport, 1997). Our central argument is that during the kind of disruptive technological innovations businesses are currently experiencing in the field of information and communication industries (Christensen, 1997; Funk, 2004), service and content providers will be better off if they choose to work closely with specialized players such as service mediaries that have the knowledge and the tools to offer seamless customer experience in the multi-service, multi-access, and multi-device world of tomorrow.

The objective of this chapter is to provide an in-depth analysis of the various elements of the telecom operators business model as the convergence of the Internet, mobile, and digital TV technologies is transforming the traditionally separate markets and businesses into a new and unpredictable business environment. This contains many new risks but also offers unprecedented business opportunities for those who learn the rules of the new game fast or, even better, are the ones who create the new rules.

In order to gain a better understanding how the role of telecom operators will evolve as we move toward a multi-channel, multi-access, and multi-service business paradigm, where customers channel surf more than ever (Nunes & Cespedes, 2003), we will analyze the current structure and future trends of multi-channel service delivery by using two frameworks: (1) the BUMMAT model (Järvelä et al., 2000; Saarinen & Tinnilä, 2002), which focuses on service delivery from an end-user perspective; and (2) the Future Value Chain framework (Hara, 1999), which describes the layered structure of the future service delivery business. By combining the perspectives of these two models, we will identify the functions, tasks, and processes that need to be in place, if an operator wants to offer a truly seamless customer experience over all available channels. Furthermore, we will also discuss who would be the right player to manage and offer the delivery of various types of services in a networked business environment.

The Future Value Chain of Service Delivery

The driving forces behind the current transformation of content and service delivery business can be crystallized into two major trends: the convergence of various technologies and the end-to-end view of the value chain. These two factors can be understood better by using the Future Value Chain framework (Hara, 1999).

The Future Value Chain framework (see Figure 3) consists of six horizontal layers. On the top are customers who want to use various types of content and applications. In order to do that, customers use different access networks ranging from mobile networks (e.g., GPRS and UMTS) to short-range wireless technologies (e.g., Bluetooth and WLAN) and all the way to digital television network access (e.g., DBV-H-based mobile datacasting). Today, the industry focus is mainly on vertical services, but in the future, the focus will shift toward a horizontal view of the services.

The third layer refers to the core network, which often today is still based on switched networks. In the future, however, the core network will be based on IP networks. The next layer consists of all the various platforms that are required in order to offer content services and other applications to end-users. The platform layer is divided into two parts: The interaction part of the platform provides services such as certification, authentication, and cryptography. The application tools, on the other hand, offer services such as conversion, content presentation, and content edition. Also in this layer, there is a profound

Figure 3. The future value chain framework (Hara, 1999)

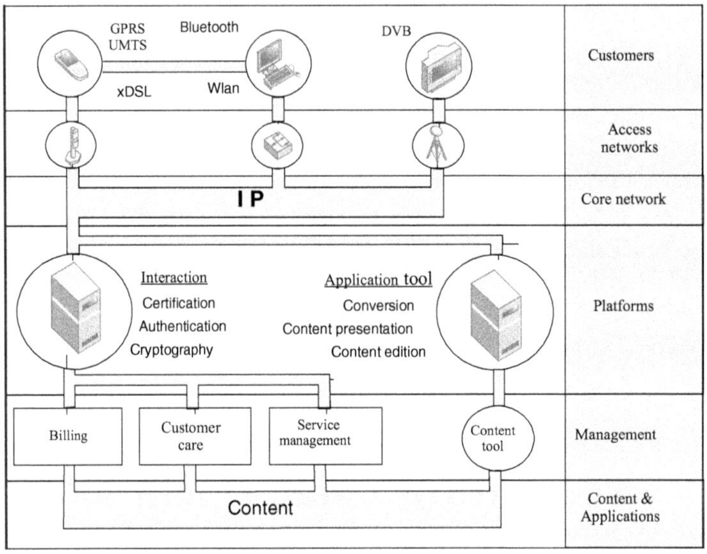

transformation taking place, as the traditional massive IN-platforms are being replaced with server-based platforms.

Layer number five is about management of content and service business. These management functions include services such as billing, customer care, service management, and, in addition, tools for content creation and management. Until now, the dominant way of taking care of management functions has been very passive, whereas, in the future, the new approach to be adopted by the leading operators is to offer interactive management tools and services to content and application providers.

Finally, the sixth layer is about content and applications. Traditionally, there has been little content available, but TeliaSonera's prediction is that, in the future, the amount of content available to end users will increase.

The Future Value Chain framework describes the various actions taking place in the service delivery value chain. However, the key questions for the operators as well as for the many players involved in this transformation process will be about an optimal position in the value chain. Table 1 summarizes the transformation that currently is taking place in the service and content delivery from the telecom operators' points of view.

Traditionally, various kinds of network services have been the main source of revenue for telecom operators. This earnings logic is currently going through a

Table 1. The transformation of the service delivery value chain

Layer	Today	Future
Access network	Vertical services	Horizontal services
Core network	Switched network	IP-based networks
Platforms	Heavy IN-platforms	Servers
Management	Passive management	Active management
Content and applications	Limited content	A lot of content

major shakeout. We argue that in the 21st century, a majority of telecom operators' revenue will come from services.

In the new service-based business paradigm, the key question will be where the core of the business is going to be located. In other words, how will the new business environment be managed and by whom, as new access technologies, content delivery mechanisms, and billing interfaces emerge in the market of multi-channel services. Likewise, there are several issues related to the content offered through wireless and wireline networks in the future: who will be delivering the content and how, and what kind of principles will be applied to the revenue sharing between the various players of the value chain.

As the technical environment becomes increasingly IP-based, the amount of peer-to-peer services and the number of content offerings increases. This will lead to a situation where nobody knows who is actually using the services or consuming the content. For operators, this will mean a choice between two very different business models: (1) "the pipe" (only the connection, using the access network and core network layers discussed earlier); or (2) manage the whole customer relationship. As seamless end-to-end user experience becomes increasingly important, the question of who, in fact, manages the whole environment will be essential for the success of the whole value chain. In the Asian mobile markets, the leading operators such as NTT DoCoMo have adopted the position of a value chain orchestrator.

Along with the evolution of mobile devices and access networks, the complexity of the business environment reaches whole new levels. At the same time, the end-users' requirements for personalization and simplicity are becoming the norm rather than the exception. It is not just the end-users who demand simplicity and ease of use; even content providers have the same requirement. We believe this goal can be reached only by adopting open standards, although we do admit that some of the biggest players in this field can force their business partners to comply with their own proprietary standards. Another benefit of using open standards in all levels of the value chain is that this approach reduces the time-to-money factor (i.e., how long it takes for a new idea to reach the market and to start generating revenue).

Although operators have been somewhat reluctant to open their platforms to third parties, we predict that this will change in the future: As the revenue no longer comes from access but from services, there is no need to be jealous about whose SIM card happens to be in the end-users' mobile phones. In this new and more open world of multi-channel content and services, the brand will be the key to success. This is why Nokia is fighting hard to maintain its strong brand visibility, even though the leading mobile operators are pushing very hard to put their own brands even on the handsets.

Finally, in order to be successful, an operator has to have a technical platform that is so flexible that it can run the future services that we are not even aware of today.

This concludes our brief introduction of the future value chain in service and content delivery business, as described by Hara (1999). Next, we will go through another, somewhat more theoretical model called BUMMAT (Järvelä & Tinnilä, 2000).

Business Model for Multi-Access Technologies

The BUMMAT (Business Model for Customer Channels Supported by Multiple Access Technologies) model facilitates identification and analysis of business models of electronic commerce as well as the related business and service concepts (see Figure 4). In addition, the model helps to identify the various players, their roles, and different ways of acting (Järvelä et al., 1999). Furthermore, the emerging business models can be described and analyzed with the help of the BUMMAT model.

The model emphasizes the fact that new access technologies and devices need to be closely integrated in the development of new business models and services. This approach is in line with the views of Lucas and Baroudi (1994), who argued that "the design of technology and design of organizations are largely becoming the same task" (p. 9). The BUMMAT model allows us to analyze the emerging access technologies (i.e., the Internet, mobile networks, digital TV, and short-range wireless networks such as Bluetooth or Wi-Fi) and the new devices (i.e., mobile phones with browsers or Java clients, interactive set-top boxes supporting MHP standard, PDAs with GPRS or WLAN connection) from different perspectives; that is, from customers' points of view, from service and content providers' points of view, and from service mediaries' points of view.

One of the major benefits of the BUMMAT model is that allows us to identify and analyze the differentiated roles that emerge in the multi-channel environment of the future. In the networked environment, actors specialize and focus on their own fields, which, in turn, increases the significance of the control of networks and service portfolios. In the multi-channel environment, there is an increasing need for coordination: Access technologies and services need to be coordinated at the technological level, and the roles and tasks of companies need to be coordinated, as well.

The BUMMAT model analyzes the structure of business at a generic level, but it lends itself also to a more detailed analysis of various viewpoints. The model highlights the numerous questions that need to be tackled in each of the four focus areas of the model (i.e., customer access, the role of service mediary, the role of content providers, and the channel operators).

The BUMMAT model is a broad concept, but in this context, we will focus on one role that is particularly interesting from the service and content delivery perspective—the Service Mediary role. According to Saarinen et al. (2002), a service mediary concentrates on developing service concepts that include the following tasks and functions:

Figure 4. The BUMMAT model (Järvela & Tinnilä, 2000, p. 108)

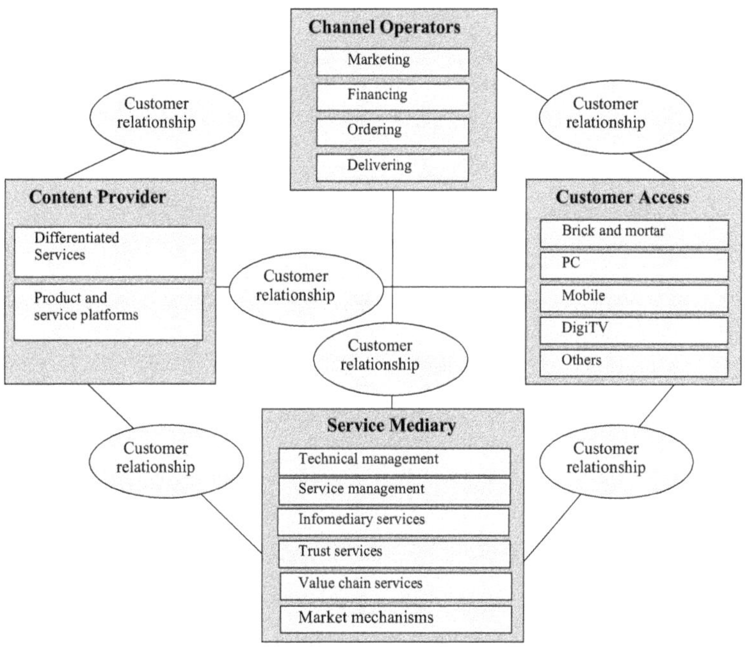

- Customer relationship management tasks consisting of gathering, analyzing, and delivering customer data
- Service management, which refers to generating road maps for new versions of services
- Technical management tasks, which include services such as detecting various types of terminals and access types, and protocols used
- Market mechanisms such as auctions

In order to better understand what the various tasks described in the BUMMAT model look like in practice, we will present examples of an earlier study where NTT DoCoMo's i-mode service was analyzed by using the BUMMAT model (Saarinen et al. 2002, 80-81) (see Table 2).

Table 2. Service mediary services offered by NTT DoCoMo (adapted from Saarinen et al., 2002)

Service Mediary Tasks	The i-mode service by NTT DoCoMo
1. Service management	
Service connectivity	DoCoMo guarantees the seamless integration of handset, network, and services.
Service concepts, variations and packaging	DoCoMo has an extensive role transaction enabler that works actively in the development of service offerings and assumes responsibility of certified services.
Service bundles	DoCoMo is involved in the decision of which service to offer in which channel.
Contract management	DoCoMo is responsible for the creation and management of service agreements.
Billing services and revenue management	DoCoMo provides billing service and revenue-sharing models to content providers.
Service and service process management	DoCoMo takes care of most business processes.
2. Infomediary services	DoCoMo manages service offerings by actively renewing and updating service agreements for the whole value network.
3. Customer relationship management (CRM)	Customer relationship is in the hands of DoCoMo; content and service providers do not have direct access to the customer base.
4. Technical management	DoCoMo supports only one access technology (however, the situation is likely to change in the future).
5. Trust services	
Identification services	Identification is not an issue as DoCoMo owns every customer.
Market maker services	The fact that a person is an i-mode user and that a service has i-mode certificate indicates that the person/service is reliable and can be trusted.
Value adding services	DoCoMo has the opportunity to build reputation or evaluation services for the i-mode user community.
6. Value chain services	DoCoMo offers a huge market with its millions of users.
7. Market mechanism services	DoCoMo has built a community of users, which makes it possible to offer auctions, portals, marketplaces, and so forth.

DoCoMo has an extensive role as transaction enabler that works actively in the development of service offerings and assumes responsibility of certified services.

This concludes our brief overview of the BUMMAT model. Next, we will combine the two models presented in this chapter in order to gain even broader understanding of how the value chain for digital content and services will evolve in the future.

The Future Model of Service Delivery

The next step on our path toward a better understanding of the structure and dynamics of the service-delivery business of the future is to develope a new framework that combines the insights of the two models presented earlier in this chapter. First we will map the various tasks, functions, and processes described in the BUMMAT model (Järvelä & Tinnilä, 2000) and the Future Value Chain (Hara, 1999). By combining the layered view of the telecom operator's business model in Figure 2 and the service mediary tasks according to the BUMMAT model presented in Table 2, we can categorize the tasks in a new way. The new categorization is presented in Table 3.

Table 3. Categorization of service mediary tasks

Layers of Business Model	Service Mediary Tasks
1. Customer perception	Service connectivity
	Service concepts, variants, and packages
	Service bundles and service management
	Personal bundles
2. Customer automation	Contract management
	Billing and revenue management
	Service and service process
	Profile management
	Service offerings
3. Technical platform	Customer relationship management
	Technical management
	Identification
	Conversion services

By taking our analysis one step further, we begin to get closer to our initial objective of this chapter, which is to identify what an optimal future model would look like. We will do this by continuing our functional analysis approach.

Service Mediary Tasks Related to Customer Perception

Our analysis indicates that an operator (or any company, for that matter) who wishes to play the role of service mediary in the service and content delivery value network of the future needs to makes sure that it possesses a rich and accurate customer perception of the existing and potential customers. First, it will have to offer a seamless connectivity by using multiple electronic and, perhaps, physical channels. Furthermore, as the operator role shifts toward an interactive role with lots of content, the service mediary can choose an active or passive role. As for the service bundles and service management, the main question is how broad a concept is offered and how this concept is delivered—by an operator or through a partner network. The fourth and final issue that is related to gaining extensive customer perception is the question of personal bundles: A service mediary integrates service offerings and offers personalized service through various channels.

Service Mediary Tasks Related to Customer Automation

The second layer in Table 3 is called *customer automation*, which consists of five different categories of service mediary tasks. The key question in the contract management category is how contract management will be organized in a multi-channel environment. The following task is related to billing and revenue management, which is also considered to be part of the customer automation layer. However, although billing typically is considered to be an operational function, revenue sharing will be a strategic issue in the future service mediary business, as the important role of a cost-efficient micro-payment system in the success of Japanese mobile Internet business indicates. As for the service and service process task, it is still unclear who will develop the services and how (i.e., how much should telecom operators get involved in the service business). Profile management becomes increasingly important in the future, as both consumers and business customers will be using more and more channels. A service mediary will be the one to maintain customer profiles for multiple channels and to distribute the profiles to service and content providers in the value network. The final service mediary task in this layer is related to building a service offering that matches the needs and wishes of the customer base.

Service Mediary Tasks Related to Technical Platform

The third layer deals with the technical platform a service mediary needs in order to handle all the tasks discussed earlier. This layer consists of customer relationship management, not just for an operator's or service mediary's own customers but for the whole value network. Furthermore, technical platform also includes the technical management of a joint service delivery platform in a multi-access, multi-channel, and multi-device environment. Identification is also one of the tasks that are taken care of by the service mediary's technical platform and offered as a service to other members of the value network. Last, but not least, the technical platform takes care of conversion services in order to guarantee a seamless customer experience, regardless of the network or terminal used.

Discussion and Conclusion

Along with the convergence of information technology, communications, and content, new kinds of challenges also surface for service providers, who will have to find ways to offer services through multiple channels, access networks, and terminals without sacrificing the quality and consistency of services offered. At the same time, as the number of channels that need to be supported increase, the services themselves become increasingly complex: In order to successfully package and deliver various elements of digital content and services, companies have to rely on external partners and must integrate various types of technologies. As a result of this development, the amount of different tasks and functions that need to be taken care of increases fast.

In order to gain a better understanding of these various tasks that emerge as we move toward the new world of multi-channel and multi-access services, we introduced two different kinds of models as the basis of a functional analysis. The Future Value Chain model (Hara, 1999) offers an overview of various technical and management functions that need to be in place, if we try to achieve a truly seamless service delivery. The BUMMAT model (Kallio et al., 1999) presented another set of tasks related to the business and customer service objectives of multi-channel service delivery. In spite of the somewhat different approaches of the two models, we managed to align the two models and thus create a broader understanding of the future requirements that the players in this field will be facing in the future.

Based on the analysis presented in this chapter, our recommendation is that more attention should be paid to the analysis of customer needs and demands. This requirement is linked to specializing roles and outsourcing in a multi-channel environment. For example, analyzing customer needs should be done by a third party, which also provides this information to other players, not as in the present system, where everyone tries to fulfill every possible role with little success. We believe that these operator-brains tasks belong to a specialized service mediary.

In this chapter, we have argued that in order to be able to offer a consistent and seamless service experience over multiple channels, service and content providers will have to team up with a new kind of player that we call a *service mediary*. We strongly believe that an individual service provider will be facing major technical challenges and cost inefficiencies, if they choose to build all the necessary systems and skills to manage service delivery in a multi-channel environment. However, in many markets, it is still unclear who will assume the role of a service mediary. As our case study of NTT DoCoMo and its i-mode service indicates (Saarinen et al., 2002), in Japan, it is the mobile operator that created the necessary services that enable successful content and service business. It is possible that operators in Europe and the U.S. will follow this example, but it may also turn out that operators in the West either prefer not to do this because of the major investments required in technical platforms (e.g., collecting and analyzing customer data; or managing content conversion, edition, and presentation) or may not be allowed to do this (e.g., regulatory and legislative issues related to the use of customer behavior data or to the bundling of access with different kinds of value-adding services). In countries such as Finland, where regulatory authorities have imposed extremely strict rules and regulations that sometimes prevent operators from developing their business toward a more comprehensive service offering, it may turn out to be difficult for operators to offer the full set of service mediary services as described in this chapter.

On the basis of the analysis of different service mediary tasks presented in this chapter, we believe that the future lies in the direction of platforms enabling business. The existing technical and service platforms will transform in the future into channels predicted in the BUMMAT model. These channels will provide the necessary infrastructure upon which enterprises easily can build their own businesses. The first signs of this development are visible, but a major restructuring awaits us as we move toward the seamless delivery of multi-channel content and services.

References

Bakos, J.Y. (1991). A strategic analysis of electronic marketplaces. *MIS Quarterly*, 295-310.

Bucklin, L.P. (1972). *Competition and evolution in the distributive trades.* Englewood Cliffs, NJ: Prentice-Hall.

Carr, N.G. (2000). Hypermediation: Commerce as clickstream. *Harvard Business Review*, 46-47.

Christensen, C. (1997). *The innovator's dilemma.* Boston: Harvard Business School Press.

Davidow, W.H., & Malone, M.S. (1992). *The virtual corporation.* New York: HarperCollins.

Dutta, A., & Roy, R. (2003). Anticipating Internet diffusion. *Communications of the ACM, 46*(2), 66-71.

Dwyer, F.R., & Welsh, M.A. (1985). Environmental relationships of the internal political economy of marketing channels. *Journal of Marketing Research,* 397-414.

English, W.D. (1985). The impact of electronic technology upon the marketing channel. *Journal of the Academy of Marketing Science, 13*(3), 57-71.

Etgar, M., & Zugman, P. (1982). The marketing intermediary as an information seller: A new approach. *Journal of Business, 55*(4), 505-515.

Fulk, J., & DeSanctis, G. (1995). Electronic communication and changing organizational forms. *Organization Science, 6*(4), 337-349.

Funk, J. (2004). Mobile disruption: The technologies and applications driving the mobile Internet. Wiley-Interscience.

Hagel III, J., & Rayport, J.F. (1997). The coming battle for customer information. *Harvard Business Review,* 53-65.

Hara, V. (1999). The future applications. *Proceedings of the Wireless Application Development*, Espoo.

Hoffman, D.L., & Novak, T.P. (1996). Marketing in hypermedia computer-mediated environments: Conceptual foundations. *Journal of Marketing, 60*, 50-68.

Järvelä, P., & Tinnilä, M. (Eds.). (2000). Elektronisesta kaupasta eLiiketoimintaan: Digitaalisen median raportti 1/2000, TEKES (in Finnish).

Kallio, J., Saarinen, T., Tinnilä, M., Tuunainen, V.K., & Vepsäläinen, A.P.J. (1999). Alternative strategies for electronic commerce—Model of customer channels. *Proceedings of the Second International Conference on Telecommunications and Electronic Commerce (ICTEC)*, Nashville, Tennessee.

Lucas Jr., H.C., & Baroudi, J. (1994). The role of information technology in organization design. *Journal of Management Information Systems, 10*, 4.

Malone, T.W. (1995). Inventing the organizations of the 21st century: Control, empowerment, and information technology. *Proceedings of the Harvard Business School Colloquium, Multimedia and the Boundaryless World.*

McNair, M.P., & May, E.G. (1978). The next revolution of the retailing wheel. *Harvard Business Review, 56*, 81-91.

Nunes, & Cespedes (2003). The customer has escaped. *Harvard Business Review*, 96-105.

Rosenbloom, B. (1999). *Marketing channels: A management view*. Dryden Press.

Saarinen, T., & Tinnilä, M. (2002). BUMMAT monikanavamalli [Research report]. Electronic Commerce Institute. LTT Research Ltd.

Shim, S., Eastlick, M., & Lotz, S. (2004). Search-purchase (S-P) strategies of multi-channel consumers: A segmentation scheme. *Journal of Marketing Channels, 11*(2/3), 33-54.

Stern, L.W., & El-Ansary, A.I. (1996). *Marketing channels*. Prentice Hall.

Stigler, G.J. (1951). The division of labor is limited by the extent of the market. *Journal of Political Economy, 59*(3), 185-193.

Tamilia, R.D., Senecal, S., & Corriveau. (2002). Conventional channels of distribution and electronic intermediaries: A functional analysis. *Journal of Marketing Channels, 9*(3/4), 27-48.

Chapter II

Developing Consumer-Preference Profiles as a Basis for Multi-Channel Service Concepts

Markku Tinnilä, Helsinki School of Economics, Finland

Anssi Öörni, Helsinki School of Economics, Finland

Anu Raijas, University of Helsinki, Finland

Abstract

Electronic distribution channels, such as the World Wide Web, mobile terminals, and digital television, have provided new opportunities for the retailing industry by providing consumers with alternative channels with which to buy goods and services. In order to better understand the implications of multi-channel access on electronic retailing, we will highlight the steps of the consumer decision-making process that rely the most on multi-channel access technologies (e.g., information search, comparison of alternatives, purchasing, and paying), describe the different channels, and use segmented groups of Finnish consumers to build sample consumer preference profiles, upon which new services can be developed.

Introduction

Electronic distribution channels, such as the World Wide Web, mobile terminals, and digital television, have provided new opportunities for the retailing industry by providing consumers with alternative channels with which to buy goods and services. These channels provide new ways to offer traditional services, such as movie ticketing, in addition to new services such as ringtones. However, despite the proliferation of e-retailing offerings, many Finnish consumers have been reluctant to adopt services across different electronic distribution channels. This suggests that either the benefits associated with one or more of these channels are not yet apparent or that the costs of switching to and using electronic distribution channels exceed the benefits. Slower adoption also may be caused partly by the shortcomings in communications technology or Web site design, including insecure or incomplete transactions, slow performance, and content or service errors. These technological deficiencies, however, will likely diminish over time, as electronic markets mature.

Instead, more attention should be paid on consumer behavior—how consumers perceive the benefits, risks, and costs of electronic channels relative to conventional ones, and which channel characteristics most affect consumer perceptions. By studying the consumer's choice of channels in a multi-channel retailing environment, retailers can build a consumer-preference profile from which new services can be more easily developed. In this chapter, we will highlight the steps of the consumer decision-making process that rely the most on multi-channel access technologies (e.g., information search, comparison of alternatives, purchasing, and paying), describe the different channels, and use segmented groups of Finnish consumers to build sample consumer-preference profiles upon which new services can be developed. The study was done by using focus-group discussions. The groups consisted of four to eight people belonging to the segments studied.

Use of Channels at Different Stages of the Buying Process

Consumer buying decisions are generally modeled as a process consisting of seven steps: need recognition, information search, evaluation of alternatives, purchase decision, purchase, use, and evaluation (see Figure 1). This model offers a generalized and comprehensive view of the consumer buying process and helps us to distinguish between the different steps of the purchase process.

Figure 1. Steps of the purchase process

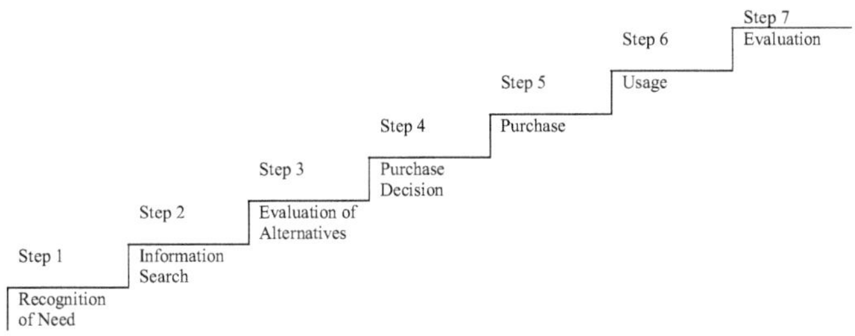

We suppose that the steps in which the use of electronic channels is most extensive are in the early stages of the purchase process, specifically information search and product comparison or evaluation.

New electronic channels provide an efficient way to provide access to information and to facilitate improved decision-making for consumers. Based on our research, these alternative channels are particularly important in the early stages of the purchasing process, where buyers are seeking information and comparison shopping. This suggests that certain goods may be sold more easily via the Internet than others, that it may make sense to market products via the Internet in support of the purchase decision, and that concentration should be on selling certain standard products first (e.g., games, music, information technology, books, and simple services). Consumers also prefer rapid, easy, and convenient access to information with capabilities for comparing alternatives because of the time and money saved. However, too much time spent searching for information could prove tedious, if users do not define their search criteria well enough to ferret out the essential information they need. Nevertheless, the final purchase is usually completed through a conventional channel. This is because consumers believe that they can bargain better in traditional channels and that electronic channels offer inadequate tools for examining the physical quality of the goods they are purchasing. Below, we highlight the steps of the consumer decision-making process that rely the heaviest on multi-channel access technologies: information search, comparison of alternatives, purchasing, and paying.

Information Retrieval

A consumer has many available sources of information regarding the items the consumer wants to purchase. The sources of information may include the consumer's own opinions and/or prior experience as well as recommendations from friends, experts, and retailers. However, a source of information is only useful for a consumer if it provides the consumer with relevant information in a timely manner. Bettman and Johnson (1991) distinguish between continuous and direct search of information. In a continuous search or update, a consumer may want to stay informed about the sellers and quality of items in the market, whereas a direct search is aimed at completing a particular purchase. The number and quality of items the consumer already owns also can influence the consumer's search for information.

In Finland, we find that consumers first become acquainted with an item they want to purchase via a conventional channel and will not use the Internet to search for information, unless they have a need to do so. This suggests that it may be preferable for retailers to make the initial customer contact via traditional means, after which a consumer can be directed to an alternative channel for additional information. Consumer groups that require the most amount of continual updates appear to be groups that have myriad purchasing needs, such as mothers with small children and young adults who want to keep abreast of market trends.

Evaluation of Alternatives

The Internet is connected closely to the buying process and is particularly helpful at the first stage of the purchase decision, when the user is becoming acquainted with the alternatives available. The role of electronic channels weakens, however, once information is obtained, as consumers try to make a purchase decision; consumers often will switch to a traditional channel to make the final purchase. For example, in the case of a customer deciding among different travel itineraries, the customer may find it easier to turn to a travel agent to schedule the best combination of flights for the best price. Possible reasons for this may include the fact that information is not updated as frequently on the Internet or that the customer wants to see and try the product before buying it. When services are bought, the complexity of the service often determines a buying place; simple services can be bought over the Internet, but when buying more complex services, consumers usually turned to experts via a more traditional channel. The only exception to this rule was purchase of less complex goods such as books and computers.

Use of electronic channels can speed up the shopping process, as the consumer becomes acquainted with the product from home, decides between products, and finds products that may not be easy to find via traditional channels. The advantages of using an electronic channel only increase as the geographical distance between the seller and buyer increases. Unfortunately, with a few notable exceptions (e.g., computer sales, house-hunting, travel services), in our research we found consumers have difficulties making product comparisons over the Internet; and, with the exception of the unique consumption habits of our younger segments, conventional channels, (e.g., newspaper ads) remain the primary sources for constant information retrieval. This could be due partly to the fact that the reliability of the information found on the Internet is low, because many retailers do not frequently update their Web sites.

Purchases and Payments

If our research is any indication, consumers prefer cash to credit as their mode of payment. Many who make their purchases via the traditional channel prefer to use cash/debit cards as an easier way to manage expenses. Some even prefer cash on delivery as opposed to paying by credit card. Via the Internet channel, however, our focus research groups only felt comfortable paying over the Internet when transactions were performed via reputable, local banks with Internet banking services. On the other hand, the consumer groups we surveyed felt positive about the prospect of payments being made via mobile phone. Many feel that, initially, only small payments should be paid using a mobile phone; however, many believe that, in the long-term, mobile payment eventually could replace cash. They emphasize, however, that mobile payments should be simple and easy to use and that widespread use of mobile payments will only become possible once advanced mobile devices have developed further and universal standards are in place, so that purchases also can be made internationally.

The Channels and Their Properties

Retail channels have diverse capabilities for provision of services. As consumer preferences often depend on the context of the purchase decision, the selection of available retail channels largely determines the consumer preferences related to choice of the channel. Furthermore, these decisions are contingent on the consumer's ability to use electronic devices, as consumers differ in their exposure and comfort level in using different channels. We will discuss four channels that are either in use or under development: the traditional channel (often known as *brick and mortar*), the Internet channel, the mobile channel, and the television and digital television channel.

The Conventional Channel

In the traditional or conventional channel, the commodities are delivered through a physical store; hence, the name *brick and mortar channel*). The advantage of the traditional channel is personal service; it is easy for the customer to confirm the reputation of the seller. For this reason, the traditional channel will

Table 1. Characteristics of the various channels

	Advantages	Disadvantages
Brick and mortar	+ personal contact + reliability of seller + seller gathers information + opportunity to try a product + familiar purchasing habits and channel + infrastructure ready	- cost of transport - opening hours restricted - limited assortment
Internet (via PC)	+ time independence + amount of information + good technical quality of voice and picture + fast comparison of products + wide additional services + wide distribution	- huge amount of information - reliable and updated information - finding a right service - safety of transactions - use of place restricted
Mobile terminal	+ time and place independence + wide penetration + familiar gadget + additional services	- small screen - slow data transmission - services expensive and inflexible - restricted content of information - no standards - integration
Digital television	+ wide penetration in the future + interactivity + quality of picture and voice + always available + easy to learn to use	- speed of adoption - consumers' doubt - technologies open - bounded use of situation?

remain an important channel for many consumer groups. In a conventional channel, sales personnel have tacit knowledge as well as familiarity with the retailer's range of offerings, all of which may be difficult to implement electronically. In the conventional channel, in face-to-face contact, it is possible to modify the service flexibly. The drawbacks to using the conventional channel include the relatively high-cost structure due to physical stores, cost of maintaining adequate work force, and limitations in opening hours.

The Internet Channel

The Internet provides consumers and businesses with new opportunities, such as time independence, a larger market, and the possibility to compare goods and services. One specific advantage of the Internet channel over conventional channels is time savings. As new electronic channels require no travel time to the store, they tend to be more efficient. Many active consumers use the Internet for communication (e.g., e-mail), information search, and bill paying. Consumers also value the Internet as one source for information when comparing the commodities, as it may provide more information than conventional means. However, despite the relatively quick adoption of the Internet for communication and conducting information searches, consumers in Finland have not widely adopted the Internet as a buying channel, despite the fact that 75% of the adult population in Finland (aged 15-74) were familiar with the Internet in 2003, and 40% use the Internet at home weekly (Taloustutkimus Ltd, 2004).

In general, our focus groups found the Internet channel especially useful when searching for dynamic information, such as checking a reservation situation or for very specific and detailed information. They also found it useful in helping to check alternatives (e.g., housing) and to save time and money, while, at the same time, avoiding the hassles associated with the traditional buying process. The drawbacks to using the Internet include the lack of reliability of the information obtained and difficulties in finding the service needed.

The Mobile Channel

In 2001, the Finnish population of 5.2 million had almost 4 million mobile phones in use (Statistics Finland, 2002). The most recent development in e-commerce is mobile devices capable of offering visually appealing and easy-to-use mobile services, which have already been in use for some years in Japan. The most significant advantages of mobile sets are place independence and widespread penetration. Mobile devices have been adopted by Finnish consumers for everyday use for communication, and retailers have had high hopes that it could serve as a new and convenient point-of-access for new services. However, usability and frequency of use for the more advanced services have proven problematic. The drawbacks have been related primarily to the development stage of the mobile phone technology, which includes the limited data transmission capability, a small screen, and relatively high usage costs.

The Digital Television Channel

Digital television is in the process of being launched in Finland. By the end of 2004, nearly 100% of Finnish viewers will have access to a digital television network. The widespread penetration of television sets, picture and voice quality, consumer familiarity with the device, and the fact that television viewing has an established role in consumers' daily routines, are the cornerstone on which the success of this new channel is built. The speed of adoption among consumers of the new television sets and services, however, are still in question. In addition, there are still open technological- and standardization-related questions. In the short run, we predict that growth will be slow. Given the fast development of technology, consumers likely will be reticent in purchasing new receivers.

Figure 2 shows how participants of the study use various channels. The arrows show what channels could be used, at least in theory, in various phases of the buying process. The figure characterizes both consumers' opportunities to conduct various activities in a certain channel and their preferences to do so.

Figure 2. The phases of the buying process in various channels

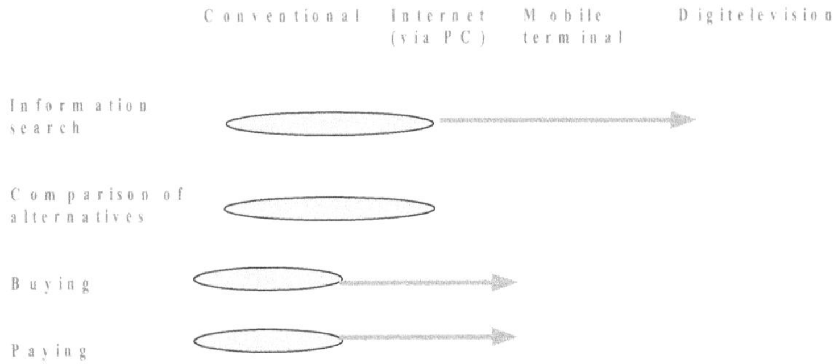

Interestingly, the traditional channel remains central to all the major consumer groups, especially in the purchase of a good or service. For retailers, this means they may need to invest more time and effort in educating their consumers on the benefits of using electronic channels as well as improving the information and services that they provide on these channels. In particular, older consumer groups with higher purchasing power likely will value being well informed about the product they want to buy. While there are many possibilities in the future for the mobile channel, it is necessary to concentrate on the services that are suitable for the channel. Absolute simplicity and clarity are required of the mobile services. Mobile commerce, such as the purchase of movie tickets, news services, and weather services could come into question only by when they are provided at a reasonable rate. This is likely when the mobile channel is the only channel possible for the service or, if by using it, time can be saved.

Benefits of Creating Consumer Profiles

Consumer profiles can help retailers to direct their services to the appropriate customer segment. For example, an Internet portal directed at retirees may reach insufficient users; however, the same portal built around the game community and directed toward youths may be successful. Consumer profiles are, by definition, generalizations; although there are differences within the groups, the purpose is to characterize each group with regard to its user habits and preferences across different channels.

Schoenbachler and Gordon (2002) have shown that channel buyer behavior is influenced by perceived risk, past purchasing experience in various channels, motivation to buy from a channel, product category, and Web site design. According to them, perceived risk includes familiarity with a company, a brand, and the Internet; the price of a product; security of information; a purchase; and a guarantee of a product. Consumers' motivations to buy from a channel is influenced by consumers' lifestyle factors, demographics, and deal proneness, as well as product availability and return policy.

The rate of adoption also may differ by consumer segment. Rogers (1995) classifies consumers into five categories, according to the speed at which they adopt new technologies and services: innovators (2.5%), early adopters (13.5%), early and late majorities (34% and 34%, respectively), and laggards (16%).

Innovators are eager to try out new ideas and are more risk-tolerant. Early adopters, while not as adventurous as innovators, are significantly influential in the social system. The early majority is eager to adopt new innovations but does not lead in their adoption, and the late majority has a skeptical and cautious attitude toward an innovation but adopt it under economic or social pressure. Laggards have the most conservative values, and their decisions are based on past experiences.

Moreover, both consumers' needs and the resources at their disposal, such as time and money, will vary at different stages of the life cycle. As a result, the demand that different consumer groups have for new services will vary by channel. For example, older groups may prefer the traditional channel, due more to force of habit, and may be less receptive to services offered via other channels. Meanwhile, having grown up with the new technologies, young adults and families may have developed a different lifestyle, one that may influence their future likeliness to adopt new services. Another example is different usage habits across various channels according to the needs of the different age groups. For example, grade school and secondary school students tend to be versatile advanced users of the Internet and access it frequently for information. For them, surfing and using search engines over the Internet are natural, familiar, and sensible tasks, and they prefer using electronic documents to printed ones.

Figure 3. Classification of innovation adoption

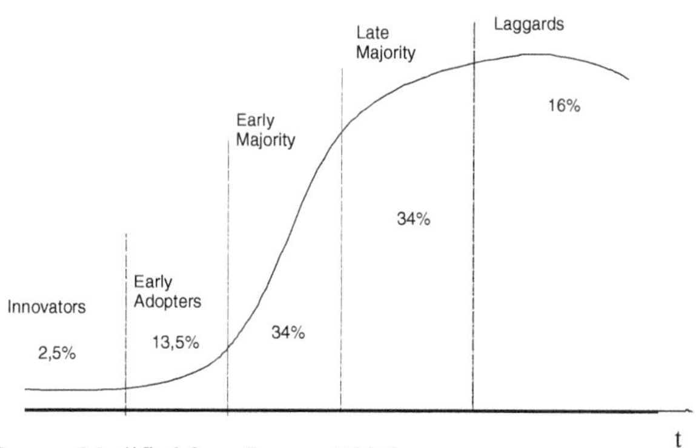

Source: Modified from Rogers (1995).

University students frequently use the Internet to aid in their studies, while mothers of small children use the Internet to obtain information on products they may want to buy. People in their 50s are occasional users of the Internet, and while they have opportunities to familiarize themselves with the Internet, their adoption of alternative channels is slow. Retirees value personal contact and are suspicious about the Internet. As a result, they have limited knowledge regarding the Internet and its use.

To shed further light on the usage patterns in these different groups, we have segmented our focus groups according to different stages in the life cycle. These six groups correspond roughly to the full range of adopters, from the innovators (children and young adults) to the laggards (retirees). These six groups are children (11 years of age and younger), secondary school students (12-17 years old), university students (21-24 years old), mothers with children in daycare/ nurseries, people in their 50s, and retirees. We describe potential retail channels based on their current behaviours and receptiveness to new ways of using technology.

Children

Children, aged 11 years old and younger, are familiar with all the channels consumers use. Most of them carry cell phones to communicate and are quite familiar with the use of the Internet. Moreover, this group is extremely influential in family buying habits, and their individual buying habits are not formulated yet.

Figure 4. Consumer research groups by stage (Source: Modified from Rogers, 1995)

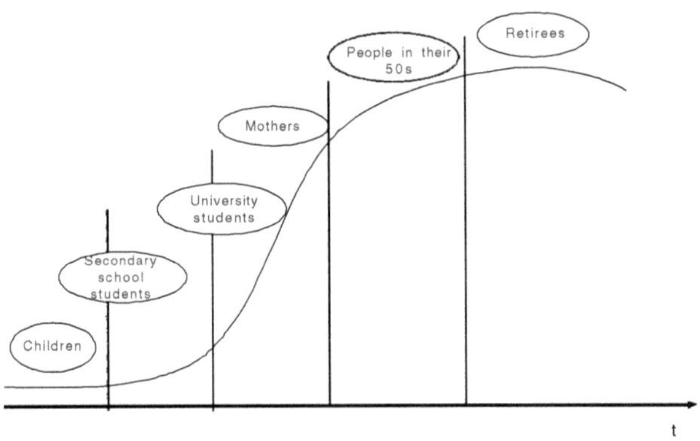

This age group's preference is orientated toward leisure (many of the children we interviewed frequently play electronic games). They also are interested especially in the purchase of games over the Internet. We assume. therefore, that this group will be comfortable using the Internet to make purchases of items besides games. From this perspective, mobile phones are perceived to be too clumsy and have poor graphics compared to game devices such as Game Boy. Nonetheless, the mobile phone is a major mode of communication for this group and, in particular, for girls who use text messages to maintain social contacts. However, there is an emphasis on the price of the mobile services. In the television and digital television channel, the demand for entertainment use is emphasized. Because the user group is extremely demanding with regard to graphics, among other things, the services of the digital television have to be distinctly better than, for example, those of the Internet, so that the channel will be used on the whole. The group we spoke with spent a fair amount of time watching television, so we predict that digital television, or *digi-TV*, as it is called in Finland, will be an interesting channel for this consumer group in the near future.

Secondary School Students

Secondary school students have grown up in the information society and appear to be quite comfortable in the use of the Internet and the mobile phone. This is due, in part, to the fact that many Finnish families own state-of-the-art technological equipment, and schools teach students how to use a computer and access the Internet. Overall Internet usage is extensive in this group, with users spending significant amounts of time surfing the Internet and using it for communication and entertainment (e.g., chatting, electronic postcards, and games). Students also use the Internet to download music and to search for information (e.g., buying computers), and many will copy music and other programs from their friends. Payment via the Internet is not a concern for this group, even though most of the members of this age group have little experience with credit card use. Instead, they routinely use Internet banks to pay for their online purchases. This group also typically spends its own money. According to statistics, the mobile channel serves as the basic communication channel for secondary school students. Although their experiences in advanced mobile services have been limited, they are not afraid to try them. Most believe mobile payment is a useful service. Similarly, we believe that television is a popular channel with this age group. This age group uses television texting often to view television schedules (in Finland, televisions equipped with special receivers also can receive broadcasted text content). Secondary school youths spend surprisingly little time watching television (an average of 1.6 hours on weekdays and 2.7

hours on the weekends). Nevertheless, we feel that there is potential for its use as an electronic channel by this group.

University Students

The purchasing power of university students tends to be more limited than the average working adult. Furthermore, university students usually have only recently set up their own households. As a result, they also tend to focus more on cost savings and inexpensive services. However, this group is the most excited about electronic channels and most frequently accesses services in different channels to prepare for purchasing decisions. Furthermore, mobility is of the utmost importance to this group, and they likely will be most attracted to services that are available to them while they are on the road.

University students utilize the Internet routinely. Indeed, a majority of the group is heavy users and even self-proclaimed addicts who view the Internet as the most abundant and most versatile source for information. The fact that Internet services are free is important. As a large number of students do not have a fixed telephone line, their access to dial-up Internet at home is limited. The mobile channel is also important to this group and is frequently used to keep in touch with personal contacts and for services. Heavy users represent as much as 76% in this group. This group is distinctly more interested than other groups in advanced mobile services and are receptive to mobile advertising as well as mobile payment, especially those requiring only small payments, and as much as 88% would prefer to conduct tasks via Internet and mobile channels only. This group requires flexibility in mobile services, especially for the sake of mobility, and looks forward to using new and advanced mobile services but wants to use services that add value instead of those that merely provide mobility. For example, they are reluctant to pay for mobile entertainment services, because they can get it from the Internet for free. Furthermore, university students express a preference for fixed payments similar to Internet access, which is offered at a flat rate in Finland. As with the secondary school students, this group also uses television texting as a substitute for viewing schedules on paper.

Mothers of Small Children

Mothers represent young families in which the home is a central focus and have high purchasing power. As caretakers of young children, the facilitation of life on a weekday is reflected in their service demands across all channels. Mothers often use traditional channels to purchase many different products and services and often in large quantities at a time. They are regular users of the Internet to obtain information and to purchase goods. We believe that, because of their focus on the home, the Internet may become a major channel for this segment in the future. They like the idea of making purchases online, because these services can save them time and hassle. In earlier studies, this group also expressed a preference for electronic grocery services. The mobile channel is an interesting channel for this group, as well, but because of the group's focus on the family, it does not have the same significance to this group as it does for young adults. Nonetheless, they value mobile services, because it allows them to remote control things at home. We also believe that mothers can become significant users of digi-TV services; however, they are constrained by their children's schedules, and this tends to affect their television viewing habits.

People in Their 50s

This group consists of aging persons whose children have left home (empty nesters). The consumption of the group is relatively low, while their purchasing power and savings rate are both relatively high. Furthermore, their buying habits are often difficult to change. For this age group, the traditional channel plays an important role, especially in more complex purchases. Consumers in this group prefer to sample the expensive products and to try them personally. Although many have access to the Internet and, in some cases, have used it to search for product information, most have relatively little experience buying goods or services over the Internet. While many use mobile phones and watch television, they neither see the possibilities for the new technology nor want to try them.

Retirees

There is a significant variance in opinion among retirees on the use of different channels. The disparity is such that a few years' difference in age can result in drastically different usage patterns. For example, recent retirees may be familiar with how to operate a PC and a mobile phone, and how to access the Internet, whereas older retirees often lack the basic know-how in these areas. Security concerns and fear of new technology may retard older retirees' receptiveness to learning to use new technologies. Consumers are comfortable primarily using the traditional channels. About 50% of this group prefers only to use the traditional channel and does not want to learn how to access any new channels (e.g., Internet banking services). Instead, this consumer group prefers to shop for goods and entertainment services, to obtain information with the assistance of services and products from the traditional channel, and to sample products before buying them. The consumption of the group is distinctly lower than those of younger families and is directed at areas such as security, health, and cleaning services (Järvelä et al., 2001). Mobile phone usage has some potential, as retirees express interest in buying services via the mobile phone, especially services related to physical safety and health. Finally, out of all the groups we spoke with, this group spends the most time watching television. On weekdays, they watch television for an average of 2.2 hours, and on weekends, an average of 3.4 hours. As a result, the group may be more comfortable with the television as a digital channel in the future. If this becomes the case, then ease of use will be important.

Recommendations for Managers

Providing access to multiple channels can be both an opportunity and a challenge for a company, as the roles of conventional and new channels are still emerging. The interaction between a consumer and a retailer seems to be important for the subjects of our study and, therefore, the importance for the facilitation of interaction in electronic channels. We also need to understand more closely the impact of electronic decision-making tools on consumers' purchasing behaviors in order to make a reasonable assessment of their effect on consumer decisions and to uncover why consumers may or may not be willing to adopt certain electronic channels for their buying decisions. Therefore, from the retailer's viewpoint, it is extremely important to (1) understand the role(s) consumers ascribe to various channels and (2) identify the determinants of various consumer choices. This knowledge is a prerequisite to providing consumers with an effective and better-integrated selection of services across multiple distribution channels. Consumer profiles can assist in this process by helping retailers direct their services to the appropriate customer segment by characterizing each group with regard to its user habits and preferences across different channels.

References

Bettman, J.R., & Johnson, E. (1991). Consumer decision making. In T.S. Robertson, & H.H. Kassarjian (Eds.), *Handbook of consumer behavior* (pp. 50-84). Englewood Cliffs, NJ: Prentice-Hall.

Järvelä, P., Raijas, A., & Tinnilä, M. (2001). *Riittävätkö kuluttajan kiinnostus ja maksuhalukkuus sähköisiin palveluihin?* Helsinki: LTT-Tutkimus Oy.

Rogers, E.M. (1995). *Diffusion of innovations*. New York: Free Press.

Schoenbachler, D., & Gordon, G. (2002). Multi-channel shopping: Understanding what drives channel choice. *The Journal of Consumer Marketing, 19*(1), 437-446.

Statistics Finland. (2002). Matkapuhelin omassa käytössä marraskuussa 2001 sukupuolen ja iän (15-74v.) mukaan. Retrieved from *http://www.tilastokeskus.fi/tk/yr/tietoyhteiskunta/matkapuhelin_oma_kuvasivu.html*

Taloustutkimus Ltd. (2004). Tuloksia Internetin käytön kehityksestä Suomessa. Retrieved from *http://www.taloustutkimus.fi/*

Tilastokeskus. (2002). Kotitalouksien laitevaranto. Retrieved from *http://www.tilastokeskus.fi/tk/yr/tietoyhteiskunta/ttyk_laitteet.html*

Tinnilä, M., & Saarinen, T. (2002). *Bummat monikanavamalli*. Helsinki: LTT-Tutkimus Oy.

Chapter III

Consumer and Merchant Adoption of Mobile Payment Solutions

Niina Mallat, Helsinki School of Economics, Finland

Tomi Dahlberg, Helsinki School of Economics, Finland

Abstract

As payments by mobile phones are an enabling technology, the adoption of mobile payments is believed to significantly influence the successful emergence of electronic and mobile commerce. In recent years, several mobile payment solutions have been launched, including the Mobipay in Spain, Moxmo in The Netherlands, M-pay in UK, and Pan-European SimPay. With the exception of mobile service purchases, however, consumer and merchant acceptance of these solutions has remained marginal. We discuss consumer and merchant adoption of mobile payments and suggest drivers and barriers for this adoption. We also describe potential mobile payment application areas and identify areas in which mobile payments have the highest and lowest possibility to succeed. The information is based on extensive research conducted among Finnish consumers and merchants. The relevance of the results to other markets also is considered briefly in the discussion section of this chapter.

Introduction

Proliferation of mobile telephony during the 1990s and the success of mobile content services, such as ringtones and logos, raised high expectations for mobile commerce. Mobile commerce is a form of electronic commerce, where at least one part of the transaction is conducted via a mobile device, mainly a mobile phone. The differences between mobile and electronic commerce lie in access device and network technologies, use experience, and use contexts. While e-commerce is conducted through stationary desktop and portable laptop computers and requires basic PC and Internet literacy, the most common devices for m-commerce are mobile phones and PDAs. These mobile devices enable transactions anytime and anywhere, require limited technical understanding, and are more personal in nature, because they commonly are used by single users, who keep the devices with them most of the time (Lyytinen & Yoo, 2002; May, 2001). On the other hand, the small size of devices and slower wireless networks limit data display, input, and transfer, as compared to e-commerce environments.

To serve the needs of the developing new mobile commerce arena, banking and telecom industries, among others, have developed mobile payment solutions. The list of applications handled by current mobile payment solutions include vending, ticketing, purchase of mobile content services (e.g., ringtones, logos, news, mobile games, etc.), electronic banking, peer-to-peer fund transfers, purchases on the Internet, and purchases of goods and services in the physical world. The most common way to make a mobile purchase is to call or send an SMS to a premium-rate service number or to send a service request to a mobile Internet site. The purchase is then charged via a monthly mobile phone bill or, in the case of prepay subscribers, deducted from a call credit. Other charging alternatives are mobile credit card billing, debiting a separate mobile account, and debiting a bank account. In Japan, the latest mobile payment solutions utilize RFID and other wireless and contactless (e.g., smart card) technologies (NE Asia online, 2004).

Since both electronic and mobile commerce environments currently lack prevailing and standardized global payment solutions, especially for micro-payments, mobile payments have the possibility to become a solution for this payment problem. For widespread acceptance and value to users, however, mobile payment solutions should be adopted in physical retailing, as well, not just on Internet and mobile networks.

To better understand mobile payment adoption, we conducted an empirical research focusing on the following three research questions: (1) Are consumers and merchants aware of mobile payment solutions? (2) What factors increase or inhibit the adoption of mobile payment solutions? (3) Which applications do consumers and merchants perceive most suitable for mobile payments? Our research draws from information systems adoption and acceptance theories, such as Diffusion of Innovations (Moore & Benbasat, 1991; Rogers, 1995) and Technology Acceptance Model (Davis, 1989; Davis et al., 1989). Empirical data were collected in 2003 among Finnish consumers and merchants with a qualitative and quantitative approach. The qualitative consumer study included focus group interviews with 46 consumers and a quantitative survey with 672 valid consumer responses. The qualitative merchant study included 15 individual merchant interviews and the quantitative study a survey with 143 valid responses. The merchants contacted represent various sectors in Finnish B2C business.

Knowledge and Experience in Mobile Payment Solutions

The adoption of a new innovation goes through a five-stage process: (1) knowledge, (2) attitude, (3) decision, (4) implementation, and (5) confirmation (Rogers, 1995). Awareness and knowledge are necessary preconditions for adoption. Therefore, the amount of knowledge people have of an innovation is an important predictor of the likelihood of adoption. Furthermore, research on technology diffusion has found that user experience in terms of trials or use of previous similar technologies is an important predictor of adoption (Agarwal & Prasad, 1999; Rogers, 1995; Taylor & Todd, 1995). We investigated both the awareness and the experiences of consumers and merchants about mobile payments and the level of knowledge the groups have about these new innovations.

Consumer Knowledge and Experience

Approximately 50% of the Finnish consumers surveyed were aware of ways to make purchases or pay bills with a mobile phone. Despite this general knowledge, however, most consumers do not know about mobile payment solutions in detail, because the solutions are just emerging, and consumers do not have enough information or personal experience about mobile payments other than mobile operators' mobile phone bills. Although consumers are interested in having more information about new payment solutions, the average level of consumer knowledge suggests that consumers are not yet informed enough to make decisions on mobile payment adoption.

Table 1. Consumer usage of mobile services and mobile payment solutions (N=672)

Purchase (have you purchased with a mobile phone)	**% of Respondents**
Ringtones, logos	50%
Other mobile content services	30%
Ticketing, vending services	10%
Payment (have you paid with a mobile phone)	11%

Consumer experiences of mobile payment solutions are still marginal, with the exception of paying for mobile content, as shown by the survey results of Table 1. These results also suggest that most consumers do not consider paying for ringtones and logos as mobile payments, since over 50% of the respondents had made such purchases, but only 11% reported that they had conducted payments with a mobile phone.

We have named the 11% of consumer respondents who have experience with mobile payments and who understand the concept better than the majority of the respondents as *early mobile payment adopters*. They are relatively younger and have higher interest in mobile technology and services. Other demographic variables, such as gender or education, were not found to be statistically significant in determining who becomes an early mobile payment adopter.

In addition to the early adopters, another interesting group of consumers is *potential users*. Over 46% of the survey respondents indicated interest in using mobile payments in the future and were identified as potential users. Compared to other respondents, on average, potential users have more education and higher professional positions, and are more interested in both technology in general and mobile technology and services in particular. Together, the early adopters and the potential users form the most likely user group for mobile payments in the future.

Merchant Knowledge and Experiences

Merchants who have participated in mobile payment pilots and tested mobile payment solutions in their businesses have the most experience with the solutions. Almost all interviewees and 46% of the survey respondents reported that they knew possibilities to offer mobile payments for their customers. Merchants are most likely to take part in pilots when they need new payment solutions, when they are interested in practical learning about a new technology, or piloting does not require too much of their resources. Our results suggest that successful pilot testing is a precondition for merchant adoption. The most likely merchant user group for mobile payments is comprised of the merchants who have participated in piloting and who find the solution suitable for their businesses.

Drivers for Mobile Payment Adoption

Innovations need to offer superior characteristics over existing solutions to be adopted. In technology adoption research, these characteristics often have been found to relate to usefulness, ease of use, compatibility with current values, experiences and needs, and positive effect on status and image (Brancheau & Wetherbe, 1990; Davis et al., 1989; Moore & Benbasat, 1991; Rogers, 1995).

Consumer Adoption Drivers

Our research suggests that several mobile payment characteristics are perceived as beneficial by a majority of survey respondents and, thus, drive the adoption of mobile payment solutions. The identified drivers are as follows:

- **Independence of Time and Place:** The possibility of making purchases anytime and anywhere was the most valued feature of mobile payments. Although this freedom may be limited in some user contexts and during some inconvenient times, the mobility factor is still the biggest asset of mobile payments when compared with other payment technologies (i.e., physical, electronic fixed line).
- **Availability:** Consumers find mobile phones a suitable payment device, because they have mobile phones with them most of the time, and, therefore, mobile payment technology is conveniently available in different situations.
- **Bypass of Queues:** Consumers consider avoiding or bypassing queues as one potential benefit of mobile payments. One example of bypassing queues is the possibility to purchase movie tickets by ordering the tickets from and into a mobile handset in advance and, thus, avoid queuing at the box office.
- **Substitution of Cash:** The possibility of reducing cash use is another benefit of mobile payments. Consumers do not always have enough cash with them, and, thus, the mobile payment option could be useful if cash were needed urgently. Lack of correct change may happen, for example, with vending machines, other coin functioning devices, and small payments in

shops and kiosks. Compared to cash, the benefits of mobile payments are that the payer always has the exact change, and there is no need to find an ATM from which to withdraw money.

A typical example of cash replacement is car parking. After parking, a person needs exact change for a car-parking meter but may lack the right kinds of coins. Most countries with high mobile phone penetration have mobile car-parking service providers such as Mint in Stockholm, Sweden (Smith, 2004). Typically, a user calls to or sends an SMS to a premium-rate service number before parking and perhaps also after parking. Sophisticated service providers debit for only the exact parking time, and the user does not need to worry about having correct change or adding coins to a car-parking meter.

Another critical situation where cash is urgently needed is when a consumer hurries to a transportation vehicle and notices that he or she does not have enough cash to pay for the fare. Helsinki city public transport in Finland, for example, has successfully offered SMS-based tram and subway tickets since 2001, and they currently sell over 50% of all single tram tickets via mobile phones. Mobile transportation ticketing services are provided in many major Asian and European cities.

In the long run, mobile payments could replace wallets, plastic, and smart cards. Consumers favor this development, because the current multitude of cards needed for different purposes is inconvenient. However, the perceived lack of safety and security of mobile devices is a potential problem in this direction of development.

Figure 1. Mobile payment characteristics perceived as important by consumers (N=672)

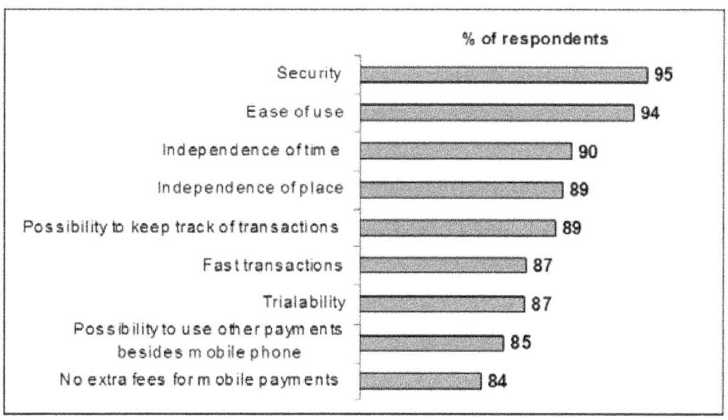

- **Economic Benefits:** Economic benefits such as price discounts, loyalty programs, and purchase bonuses are lucrative for consumers. These economic benefits could increase consumer interest in mobile payments, especially during service launch.
- **Other Benefits Consumers Require:** Currently used payment instruments (typically debit cards, credit cards, and cash) are easy and fast to use, inexpensive, widely accepted, and secure. Moreover, cash is currently the only payment method that allows anonymous payments and person-to-person money transfers, (e.g., lending money to a friend). To be able to compete with the dominant payment instruments and to fulfill consumer expectations, mobile payments should offer these benefits, too. Figure 1 presents the factors consumers consider most important in mobile payment solutions, according to our survey.

Merchant Adoption Drivers

Merchants benefit from the new mobile payment solutions, if they are able to offer new services, increase sales, or reduce costs. These benefits are dependent on the market penetration and costs of implementing and using the solution. To ensure widespread market penetration, merchants are interested in solutions that add value to consumers and are likely to be adopted. Given that the pricing of mobile payment solutions is competitive and that many consumers use these solutions, interviewed merchants identified several possibilities to offer new services and products, to enhance existing ones, to increase sales, and to reduce costs.

- **Increased Impulse Purchases:** Mobile payments enable consumers to buy more things on impulse and, therefore, have the potential to increase merchant sales. For example, there may be an advertisement in a newspaper, on TV, or on the radio, along with instructions on making a purchase with a mobile phone. The possibility of the purchase increases when consumers can make the purchase instantly after receiving an impulse.
- **New or Better Services:** Mobile payments enable merchants to offer new or better services to customers. Especially the micropayment price category includes products and services that are impossible to offer with current payment instruments. An efficient mobile payment solution could inspire merchants to create new, additional, or enhanced services that would increase their overall sales.
- **Enhanced Product and Service Availability:** Mobile payments have the capacity of enhancing the availability of products and services and of

creating an additional order and delivery channel. Purchasing becomes independent of time and place and more convenient to customers. Especially digital products that can be transmitted directly to customers' mobile phones benefit from the enhanced remote purchase possibilities. Transport for London, for example, offers customers a possibility to pay for the congestion charge by mobile phone using an SMS text message. In early 2004, 20% of the payments were handled via text messages, and their usage continues to increase at the expense of retail and call-center channels (TfL, 2004).

- **New Customers:** Mobile payments may attract new customers, especially if purchases become easier and more available. Merchants estimate that mobile payments most likely attract young people who are experienced mobile phone users, teenagers who do not have debit cards, individuals who do not carry much cash with them, and technologically oriented persons who may prefer the use of mobile phones to other payment alternatives. McDonald's Slovenia, for example, recently teamed up with the local Mobitel mobile network operator to better serve those customers who prefer non-cash payments (Mobitel, 2004).
- **Improved Company Image:** For some merchants, mobile payments may improve their companies' images. Consumers regard them as innovative forerunners in their industries. Merchant opinions on the image are somewhat contradictory, however, and some perceive that the solutions are currently too immature to have a positive effect on company image.
- **Lower Commissions and Costs:** Merchants would benefit if mobile payment service providers charged lower commissions than the commissions charged on competing payment alternatives (e.g., credit cards). At least, the costs of mobile payments should equal the costs of competing alternatives.
- **Faster Settlements:** Similar to lower costs, payment settlement times should be faster than or equal to those of competing alternatives in order for mobile payments to be competitive. Only if single payment clearances are very small may merchants prefer the bundling, clearing, and netting of payments into larger sums.
- **Improved Efficiency:** Mobile payments could reduce costs, if they improve the efficiency of payment process or reduce employer work on payment processing. One example of improved efficiency would be the transformation of manned service stations to unmanned self service stations during off-peak hours.

Barriers to Mobile Payment Adoption

In addition to adoption drivers, innovations also face adoption barriers. Barriers are complexities in understanding and using the innovation, costs of adoption and use, and perceived risks (Davis et al., 1989; Moore & Benbasat, 1991; Rogers, 1995). Recent research in electronic and mobile commerce has discovered that trust and security play a significant role in the adoption of these new forms of commerce (Gefen et al., 2003; Jarvenpaa et al., 2000; McKnight et al., 2002; Pavlou, 2003).

Consumer Adoption Barriers

Several factors currently limit consumer adoption of mobile payments. The following barriers were identified in our empirical study, starting from the most common.

- **Premium Pricing:** Many consumers abstain from paying with mobile phones because of premium pricing. Consumers are especially reluctant to pay separate fees for mobile payment services. Bundling mobile payment fees with other financial services would be a more acceptable solution. The perceived advantages of the mobile payment service will tend to increase the willingness to pay for the service.
- **Limited Possibilities to Use:** Lack of widespread merchant acceptance prevents consumer adoption. Mobile payments are currently a marginal payment technology and are not available or accepted widely enough. Mobile payments should become more common in daily purchase situations so that consumers would learn the new payment instrument and become accustomed to using it.
- **Lack of Convenient Billing Solutions:** A majority of consumers perceive current mobile payment billing solutions as difficult. Separate accounts require inconvenient preparations before use, including registration and opening the account. During use, it is difficult to keep track of the balance of the account and to transfer money to and from the account. Mobile phone or credit card billing may end in an unexpectedly large bill at the end of a month. Finally, bank account statements may be littered with small account entries, if a bank account is used to debit numerous small value mobile payments.

While billing preferences may vary from country to country, our results suggest that mobile payment solutions should be easy to use and compatible with existing payment instruments and financial services. Lack of compatible billing procedures seems to reduce or postpone consumer adoption of mobile payment solutions.

- **Trust and Security:** Finnish consumers regard banks, credit card companies, and telecom operators as reliable mobile payment service providers. Of these enterprises, banks are perceived as slightly more reliable than others, probably because Finnish consumers have learned to trust banks due to their long customer relationships with banks. In general, the results indicate that, in mobile payment services, consumers are likely to rely on traditional financial institutions, which they also have learned to use for their other financial affairs. Telecom operators are perceived as reliable, because consumers expect them to have the knowledge and skills about the new mobile technology. Small and unknown payment service providers, including startups, were perceived as unreliable by consumers.

Despite the general reliability of the dominant mobile payment service providers, consumers perceived several risks in using mobile payments. The perceived risks identified in our study fall into the following six categories:

1. Unauthorized use of the payment instrument
2. Transaction errors
3. Lack of transaction record and documentation
4. Vagueness of the transaction verification
5. Privacy concerns
6. Device and mobile network reliability

To reduce perceived risks, mobile payment service providers should ensure that the payment infrastructure is secure and reliable. They also need to communicate the terms and policies of the service to consumers and provide enough feedback and documentation during the use of the service. Transaction errors, unauthorized use, and the like can be prevented with certificates such as security PIN codes (e.g., a WPKI digital signature). There is, however, a tradeoff between security and ease of use. Although certification techniques increase the security of the service, they also make it slower and more difficult to use. It is the challenging task of service providers to find the golden mean between security and ease of use.

- **Lack of Information:** Many consumers are interested in mobile payments but have little information. Perceived trust and security risks clearly indicate this. As mobile payments are not yet widely used and, therefore, not visible to consumers, better methods to spread information about the new payment services are needed.
- **Difficulty of Use:** Consumers expect that mobile payment solutions should be easy to use, whereas current experiences suggest the opposite. Difficulties lie both in the introduction phase of the new payment technology and during continuous use. The introduction phase may appear difficult, if the mobile payment service requires complex registration procedures and separate billing arrangements. During continuous use, the payment procedure should be simple (i.e., press a single key) and fast. If payments are conducted with SMS messages or mobile Internet browsing, however, the message formats often are complicated and slow to key in, and the various codes and premium service numbers or sites are difficult to remember. It is also difficult to keep track of the conducted mobile payments due to lack of documentation and to find instructions to conduct mobile transactions, not to mention charge-back situations.
- **Sufficient Existing Payment Instruments:** Current payment instruments are widely accepted and cover a majority of purchasing situations. With the exception of lack of cash and problems encountered in some special occasions, there are few situations in the physical world where new payment instruments are clearly needed. Further, as purchases on the Internet and via mobile networks are still infrequent, the lack of mobile payment instruments in networks does not appear to be a significant disadvantage for the majority of consumers. In the survey study, only 5% of the respondents indicated that they would consider mobile payments as their primary payment technology. This result suggests that existing payment methods are sufficient for the majority of the respondents most of the time.
- **Employer-Paid Mobile Phone Bills:** Consumers whose mobile phone usage and subscription fees are paid by their employers may have limited access to mobile purchases and payment services due to restrictions set by their employers.

Adoption Barriers for Early Adopters and Potential Users

Early adopters and potential users find the adoption barriers significantly lower than other consumers. Compared to consumers in general, early adopters have a better understanding of mobile payments, show more trust in mobile payment service providers and technology, perceive fewer risks in payments, and consider mobile payments easier to learn and use. Service cost, however, is a critical barrier for early adopters and potential users as well. To conclude, pricing is an especially important mobile payment adoption factor.

Merchant Adoption Barriers

Similar to consumers, merchants also experience barriers for mobile payment adoption. The following are barriers identified in our empirical study.

- **Incompatibility with Business Practices:** One significant barrier to the merchant adoption of mobile payments is the perceived incompatibility of mobile payment technology with existing business practices. For example, while mobile payments are especially suitable for digital content, their applicability in supermarket checkout, is less obvious. Merchants' businesses largely determine whether they find mobile payments compatible or not.

  Further, many merchants feel that the current mobile payment solutions have been developed to serve the payment service providers' interests, whereas the benefits to merchants and consumers have not been emphasized.
- **Lack of Users and Use:** Another essential problem with mobile payments is that too few consumers currently use this payment technology. Only a large amount of users makes it profitable for merchants to offer payment with a new instrument. For some merchants, widespread adoption is acceptable in the long run, whereas others require high usage rates before they adopt mobile payment solutions.
- **Investment and Usage Costs:** Similar to consumer adoption, the high costs of mobile payment solutions also prevent merchant adoption. In particular, the commissions of mobile payment service providers were perceived to be too high, even to the point that it becomes unprofitable for merchants to offer certain mobile services.

  Cost-effectiveness and competitive pricing are prerequisites for mobile payment adoption. Commissions should be competitive with or lower than the commissions of the current payment methods. Further, mobile payment service providers should support the early adoption of mobile payments with reduced fees or free-of-charge trials until the solution gains widespread acceptance and use. Currently, merchants see little or no cost advantages in mobile payment solutions.

- **Difficulty of Use:** Merchants are concerned about the usability of mobile payment solutions, which they see as an essential factor in consumer adoption. These solutions become commonly used, if they are simple and fast to use. Many merchants perceive current solutions as cumbersome and slow to use. Especially SMS messages and separate mobile payment accounts are seen in this light. Speed is considered important by retailers as well as by many digital service providers.
- **Lack of Standards:** Lack of standards between new mobile payment solutions is another barrier for adoption. Merchants expressed three generic standardization requirements for mobile payments. First, mobile payment solutions need to be independent of single banks, telecom operators, and devices in order to be usable for a large variety of consumers. Second, they need to be compatible with cash register and electronic payment technology (standards) at point of sale. Third, mobile payment solutions should support mobile commerce and payment transaction roaming between service providers, because there is little room in the market for several new, competing, and incompatible payment standards. Different solutions may confuse consumers and make it too costly for merchants to provide services for all the different payment standards. The worst outcome is that both merchants and consumers become disinterested, and all new solutions remain unused. Cooperation and standardization between payment service providers is, therefore, recommended.
- **Trust and Security:** Trust in payment service providers and security of payment solutions are other prerequisites for merchant adoption of mobile payments. The surveyed merchants considered mobile payment solutions as fairly secure. Similar to consumers, Finnish merchants perceive banks and mobile operators as reliable mobile payment service providers. However, approximately one-half relied more on banks in security issues, including certification.

Precise and timely settlements are an important means of enhancing merchant trust in mobile payments. Verification and certification during payment transaction processing confirms that the payment has been successful and that the payee's account is credited. The reliability of the payment infrastructure in all situations is another critical security issue, as reliability prevents errors, fraud, and lost sales.

Preferred Applications

Our results indicate that both consumers and merchants have clear ideas about suitable applications when they are asked to consider the use of mobile payments.

Applications Preferred by Consumers

Our findings suggest that consumers are most interested in using mobile payments in small value purchases, where mobile payments are used instead of cash instruments or to pay for digital content. In the survey, we asked consumers to list those purchases for which they would be most interested in paying with a mobile phone. Table 2 shows that the most often selected applications for mobile payments are mobile content and services, mobile banking and money transfer, and payments related to various types of travel. The interest in paying for the listed items with a mobile phone is significantly larger than those currently using mobile phones to pay for these services. It is possible that this finding relates more to expected removal of barriers than to drivers. Although the indicated interest does not necessarily result in future use, it is a good sign for the latent use potential among consumers.

Table 2. Consumer interest toward mobile payment applications (N=672)

Purchase/Payment	% of Respondents
Ringtones, logos	57%
Mobile content such as weather, news, directory, route information, or games	55%
Car parking	54%
Ticketing, such as movie or concert tickets	52%
Mobile banking, such as bill payment	52%
Money transfer from person to person	51%
Taxi	44%
Car wash, refueling, and other car services	42%
Long-distance traffic ticket	41%
Vending machines for drinks, candies, passport photos, video rental	39%
Local transportation fares	38%
Hotels, accommodations	37%
Betting, lottery, gaming	36%
Lockers, storage rooms	35%
Lunch, fast food	33%
Restaurants	33%
Airline ticket purchases via the Internet	31%
Groceries	31%
Purchases in kiosks	28%
Internet purchases of books, clothing, music, games, films	27%
Consumer durables, such as electronics, CDs, videos	23%
Purchases on TV shop and digital-TV	21%
Dealing in shares and securities, related mobile services	16%

Table 3. Suitable products and services for mobile payments listed by the interviewees

Product/Service
Ticketing (e.g., events, travel, entrances, movies)
Vending machine purchases (e.g., drinks, candies, newspapers)
Digital and mobile content and services (e.g., news, weather, games, music)
Parking
Self-services, purchases from unmanned service stations (e.g., carwashes, travel insurances)
Remote payments, subscriptions, home deliveries

Merchant Preferred Applications

Merchants regard mobile payments as most suitable for small value purchases of digital content and services, because they can be sent directly to a mobile phone. Ticketing is especially seen as a potential service. Also, mobile content purchases, vending and parking services, remote payments, and self-services in general were seen to be well suited for mobile payments. Table 3 shows the most suitable applications for mobile payments, as listed by the interviewed merchants.

In the merchant survey, the respondents were asked to choose products and services that are best suited for mobile payments in the future from a list similar to that in the consumer survey. Of the 143 respondents, 19 selected Internet purchases, 17 daily consumer goods, 13 lunch and fast food, and 11 consumer durables. The relatively high figures of retail and restaurant purchases may be explained by the industries of the respondents: 28% of respondents represented restaurants and 17% stores and supermarkets.

The results indicate that a majority of the merchants have opinions similar to consumers and see mobile payments to best suit purchases of mobile content, ticketing, vending, and information networks. A small minority of retailers were willing to try mobile payments in physical point of sale, but the potential of mobile payment applications still lies in the digital goods and services and in small-value, real-world payments.

Concluding Remarks

The recently introduced new electronic payment technology—mobile payment—has come boldly to the market to conquer its place among established payment instruments such as cash, credit cards, and debit cards. Since it is still in its infancy, the adoption rates of the new payment technology have been modest, and, for the most part, mobile phones are used to pay for mobile services and content. This chapter has discussed consumer and merchant adoption of mobile payments in terms of awareness, adoption drivers and barriers, and preferred applications, based on empirical evidence provided by consumer and merchant research.

The word *curiosity* best characterizes current consumer and merchant awareness of mobile payments. While mobile payment service providers have piloted their solutions, often with low profiles, more efficient marketing is needed to bring the new services to the consciousness of intended users. Consumers need information about the use and benefits of these solutions, and merchants need information about the advantages in relation to their specific business.

Consumer adoption of mobile payment technology is driven by independence of time and place, the possibility to avoid queues, the ability to substitute cash, and envisioned economic benefits. Yet, mobile payments also should offer ease of use, widespread merchant acceptance, speed, and security matching that of the traditional payment technologies and instruments.

Merchant adoption drivers include sales increases and cost reductions. Sales increases take place through increased impulse purchases, easier or faster payments, new customers, or enhanced opportunities to sell products and services. Cost reductions may result from lower commissions or payment processing costs, more efficient payment processes, or faster settlement of payments. Some merchants and service providers highlight the importance of positive image effects, as well.

The primary barriers for consumer mobile payment adoption are premium pricing, cumbersome SMS interface, limited merchant acceptance, perceived risks, and satisfaction with existing payment instruments. Correspondingly, merchant barriers include uncertainty about the profitability and other benefits of mobile payments, low penetration rates of mobile payments among consumers, and non-standardized solutions with many payment service providers and competing technologies. Some merchants also perceive mobile payments as incompatible with their business.

There are three applications that are seen to best suit mobile payments. First, in most situations, a mobile phone is the most convenient and sometimes the only possible payment technology for mobile content and service purchases. Therefore, we expect that mobile payments will continue to be used in mobile and digital services, including content such as electronic ticketing. Second, the diminishing use of cash in the physical world makes it important to develop new compensatory payment instruments for small value purchases on automated machines, cash desks, and self-service stations. Mobile payments have the opportunity to provide efficient payment instruments to complement cash payments. Finally, there is a need for a cost effective means to charge small value payments in remote electronic commerce. Mobile payment instruments are suitable candidates in this area, as well.

Are our findings relevant to other markets? As our empirical findings are based on a sample on a single country, we offer the following general suggestions derived from our experience, from international contacts with experts in the field, and from literature review.

Payment cultures, instruments, jurisdictions, and infrastructures differ significantly between nations. For example, the U.S. payment system is heavily characterized by check use; Japan relies on cash; and most European countries rely on bank giro or postal giro payments, all with significantly different local payment infrastructures (Hancock & Humphrey, 1998). Another example concerning regional regulation comes from the EU jurisdiction, where a limited credit institution concept was introduced recently to allow limited credited commerce for third-party services and products. This makes it possible for mobile operators to offer third-party mobile services via an operator's network and charge for the services with an operator's phone bill without acquiring a banking license, which was required previously (T2R Final Report, 2003).

Caution with the generalization of our findings is necessary, since mobile payment instruments introduced to specific markets most likely will differ and integrate regionally or globally only gradually with overall economic integration and international network roaming. If mobile operators are able to establish international mobile payment transaction roaming similar to international mobile voice roaming, local mobile payment solutions could become internationally accessible. As has happened with physical and fixed-line electronic payments, mobile credit card instruments supported by local issuers have a high probability to succeed. Consistency with local payment culture, jurisdiction, existing payment infrastructure, and instruments is a prerequisite for the adoption of mobile payments. On the other hand, we believe that within the context of cultural, jurisdictional, and infrastructure differences, the fundamental factors that make a new technology appealing to consumers and merchants are most likely similar, although some specific adoption drivers and barriers would differ. We expect this to be true, especially in the context of mobile service purchases.

Interest in the future use of mobile payments is notably higher than the current use (A.T. Kearney, 2002). Although actual future use cannot be predicted directly from indications of interest, research results suggest that intended users are willing to consider the adoption of this new payment technology. To make new payment services successful, mobile payment service providers need to ensure that the advantages of adoption drivers exceed the disadvantages of barriers and that the new payment technology is competitive with established traditional payment technologies.

In the short run, mobile payments are not likely to substitute traditional payment technologies but will be used in digital environments and in the physical world, when cash is not available, payment cards are not accepted, or it is inconvenient to pay with a payment card. In the long run, it is possible that payment cards will be integrated into mobile devices. This would result in a more profound change in payment instruments and would require years to diffuse. Before that, mobile payment solutions need to overcome their current adoption barriers and establish a solid reputation and a position as a viable payment technology.

References

Agarwal, R., & Prasad, J. (1999). Are individual differences germane to the acceptance of new information technologies? *Decision Sciences, 30*(2), 361-391.

A.T. Kearney & Judge Institute of Management, Cambridge University. (2002). Mobinet 4 study. Retrieved October 4, 2004, from *http://www.atkearney.com/main.taf?p=5,4,1,50*

Brancheau, J.C., & Wetherbe, J.C. (1990). The adoption of spreadsheet software: Testing innovation diffusion theory in the context of end-user computing. *Information Systems Research, 1*(2), 115-143.

Davis, F. (1989). Perceived usefulness, perceived ease of use, and user acceptance of information technology. *MIS Quarterly, 13*(3) 319-339.

Davis, F., Bagozzi, R., & Warshaw, P. (1989). User acceptance of computer technology: A comparison of two theoretical models. *Management Science, 35*(8), 982-1002.

Gefen, D., Karahanna, E., & Straub, D.W. (2003). Trust and TAM in online shopping: An integrated model. *MIS Quarterly, 27*(1), 51-90.

Hancock, D., & Humphrey, D.B. (1998). Payment transactions, instruments, and systems: A survey. *Journal of Banking & Finance, 21*, 1573-1624.

Jarvenpaa, S.L., Tractinsky, N., & Vitale, M. (2000). Consumer trust in an Internet store. *Information Technology and Management. 1*(1-2), 45-71.

Lyytinen, K., & Yoo, Y. (2002). Research commentary: The next wave of nomadic computing. *Information Systems Research, 13*(4), 377-388.

May, P. (2001). Mobile commerce—Opportunities, applications, and technologies of wireless business. Cambridge University Press.

McKnight, H.D., Choudhury, V., & Kacmar, C. (2002). Developing and validating trust measures for e-commerce: An integrative typology. *Information Systems Research, 13*(13), 334-359.

Mobitel. (2004). Paying with moneta also in McDonald's. Retrieved October 4, 2004, from *http://www.mobitel.si/eng/*

Moore, G.C., & Benbasat, I. (1991). Development of an instrument to measure the perceptions of adopting an information technology innovation. *Information Systems Research, 2*(3), 192-222.

NE Asia Online. (2004). NTT DoCoMo to debut i-mode smart-card handsets capable of making electronic payments. Retrieved October 4, 2004, from *http://neasia.nikkeibp.com/wcs/leaf/CID/onair/asabt/news/314459*

Pavlou, P.A. (2003). Consumer acceptance of electronic commerce: Integrating trust and risk with the technology acceptance model. *International Journal of Electronic Commerce, 7*(3), 101-134.

Rogers, E.M. (1995). *Diffusion of innovations*. New York: Free Press.

Smith, B. (2004). Time's up for parking meters. *Wireless Week, 10*(13), 24.

T2R Final Report. (2003). Trusted transaction roaming EU project final report. Retrieved October 4, 2004, from *http://www.radicchio.cc/home/t2r.htm*

Taylor, S., & Todd, P.A. (1995). Assessing IT usage: The role of prior experience. *MIS Quarterly, 19*(4), 561-570.

Transport for London. (2004). Congestion charging central London, impacts monitoring, second annual report. Retrieved October 4, 2004, from *http://www.tfl.gov.uk*

Chapter IV

Objectives of Search and Combination of Information Channels in Electronic Consumer Markets: An Explorative Study

Anssi Öörni, Helsinki School of Economics, Finland

Abstract

Internet-based commerce is expected to radically affect many consumer markets. Present knowledge of the forthcoming transformations is based largely on the principle of value maximization; in standard economic theories of search, consumers are assumed to be both able and motivated to rationally weigh the costs and benefits they expect to ensue from search. High search costs are offered to explain why consumers evidently do not fully inform themselves about the available options. Electronic media are expected to lower search costs and, thus, radically enhance consumer search. However, little empirical evidence exists in support of these assumptions. The combination of information sources consumers prefer to use is largely unexplored in relation to electronic information sources. Likewise, our knowledge of the objectives related to the procurement decision that leads consumers to adopt electronic sources for product information is still lacking. In this chapter, we examine the determinants of using Internet-based information sources, particularly the World Wide Web.

Introduction

The existing literature on consumer search in electronic markets focuses on the hypothesized increase in accuracy and the extent of product comparison leading to enhanced market functioning. Alba et al. (1997) propose that interactive home shopping (IHS) environments provide consumers with easy access to a radically greater number of alternatives and, thus, may help them to make better procurement decisions. Bakos (1991, 1997, 1998) argues that electronic consumer markets will facilitate product comparison to the extent that the ensuing price competition will render electronic markets more efficient than their conventional counterparts. Whinston et al. (1997) propose that simultaneous, rather than sequential, search strategies may enable buyers to execute more efficient product information searches on the Internet.

Accuracy of consumer decisions has been central in previous academic efforts that focused on product information search. Thus, the benefits of electronic markets have been largely defined in a decision space of at least two dimensions: price and product attributes. Tools such as comparison matrices (Haubl & Trifts, 2000) and software agents (Vulkan, 1999) are being developed to assist buyers in product comparison. Despite these developments, there has been little empirical evidence in support of the hypothesized positive effects of electronic markets. Brynjolfsson and Smith (2000) find substantial price dispersion in electronic markets for books and CDs. Clemons et al. (1999) observe a similar pattern in the electronic markets for US domestic flight seats. Lee (1998) detects higher prices for second-hand cars in electronic markets than in conventional markets. Examination of DealPilot (later EvenBetter and DealTime), a software comparison tool (Brynjolfsson & Smith, 1999), indicates that even price-conscious buyers factor other attributes beside price into their buying decisions. Furthermore, Brynjolfsson and Smith (2000) propose that the service component (e.g., logistics) is inherent in electronic retailing and may enable sellers to differentiate even commodity products.

In conclusion, it could be argued that, to date, electronic markets have had little observable impact on the efficiency of consumer decisions. However, consumer decision making involves more than just solving procurement and consumption problems. Bettman et al. (1998) propose that construction of consumer decisions may, in fact, constitute more of the search process than decision making. They attribute the difficulty of decision making, to a substantial degree, to the limited human processing capacity resulting in lack of well-defined preferences. Rather, consumers construct preferences using a variety of strategies contingent on the demands of the task. Bettman, et al. grounded their insight on the bounded rationality put forth by Simon (1955, 1982). The following quote captures the essence of bounded rationality.

> *The capacity of the human mind for formulating and solving complex problems is very small compared with the size of the problems whose solution is required for objectively rational behavior in the real world (Simon, 1957, p. 198).*

Thus, not only are human cognitive abilities stretched while solving problems, but also problem formulation may exert excessive mental burden. Consumers are known to apply a variety of decision strategies, the selection of which depends on the individual, the task, and the environment (Bettman et al., 1990). We propose that the positive effects of using electronic information sources may accrue from sources not well captured by the value maximization framework. In particular, Web-based retailers provide information that consumers can combine with information retrieved from other sources. However, relatively little is known about how consumers combine the electronic and conventional information sources available to them in their decision environment. Of particular interest should be the motives that guide consumers in their selection of information sources. Recognition of the strong and weak points of electronic media in relation to conventional information sources would facilitate development of electronic services and should lead sellers to wield a more efficient combination of marketing channels.

The aim of this study is to describe the role of the World Wide Web (WWW) in consumer information search in travel services context. In particular, we explore the combinations of information sources, either conventional or electronic, that consumers use while searching for travel services. Furthermore, we examine how decision criteria affect the role of the WWW as an information source. To reach our aim, we first review the literature related to information search in electronic markets and consumer search, in general. Of particular interest are the benefits previously related in the literature to consumer search on the Internet. In the third chapter, we present the data collection method used, evaluate the data, and discuss the selected analytical methods. The fourth chapter contains the description of the analyses and the results. In the last chapter, we discuss the results and the limitations of the study and provide some pointers for future research.

Benefits of Electronic Markets to Consumers

In the past literature, several claims have been made concerning the possible benefits of electronic consumer markets. It is hypothesized that a greater number of alternatives will enable consumers to more accurately select the goods or services that best match their needs and wants (Alba et al., 1997). The low cost of information retrieval is expected to enable consumers to extend search for price information and product attributes and to endorse rational product comparison (Bakos, 1991b, 1997, 1998). Easy price comparison should lead consumers becoming increasingly price sensitive. However, Alba et al. (1997) and Bakos (1997) both note that electronic markets do not facilitate retrieval of price information only. Independently, the cost of search for product attribute information also could be lowered, which should lead to a decrease in price sensitivity. Electronic markets also are expected to dilute the importance of store location (Balasubramanian, 1998; Rayport & Sviokla, 1994). Past research (Eastlick & Feinberg 1994) has shown that convenience is a principal reason for patronage of direct channels. Eastlick and Feinberg (1994) argue that retail stores provide many benefits over direct channels, such as quicker gratification, opportunity for physical inspection, and easier product returns. According to them, in the absence of transportation costs, consumers typically would purchase from the retailer rather than the direct channel. Thus, Eastlick and Feinberg (1994) distinguish between merchandise-related and search-process-related factors. Simon (1987) indicates the same difference, while accusing economics of focusing on outcomes of consumer decisions and neglecting the decision processes. Even if theoretically sound, the tenet of consumers advancing product-related rather than process-related objectives while using electronic markets has received little empirical support. However, empirical evidence exists to support the relationship between the objective of purchase convenience and the use of direct channels. We propose that, in this respect, electronic media are akin to other direct sales channels rather than being different from any other source.

Proposition 1

Consumers using electronic information sources are likely to differ from consumers not using electronic sources in their attempt to advance objectives related to the purchase process (i.e., convenience) rather than those related to the product or merchandise. A corollary of this proposition is that service- or merchandise-related factors gain importance, only if search costs are driven very low.

Consumer search strategy is found to be associated with the combination of information sources used (Fodness & Murray, 1998). Thus, the objectives of search are likely to determine, to a degree, the use of electronic information sources. Price and quality are objectives that are related to electronic sources almost ritualistically (Bakos, 1991b, 1997, 1998). Electronic sources are touted to posses qualities not related to more conventional sources; that is, electronic sources are expected to be radically more efficient in information dissemination, and, thus, the related search costs are very low as opposed to what has been observed in relation to other sources. If this hypothesis holds, consumers should be able to find what they are seeking by using electronic sources only. However, we have observed in relation to our previous experiments that information retrieval often is very difficult using electronic sources only (Korhonen et al., 2000).

Proposition 2

Current electronic sources do not facilitate comparison of travel services, as is hypothesized. Rather than being used independently, it is plausible that electronic sources are used to complement other information sources. A corollary of this proposition is that electronic sources share the consumer objectives with those sources they typically complement.

Tourist Information Search

Extensive literature on tourist information search exists (Snepenger & Snepenger 1993), relating the correlates of search effort to search strategies and information sources used. Information acquisition often is necessary for decisions preceding a vacation, such as selecting a destination, and also for on-site decisions such as selecting accommodations, transportation, and tours (Filiatrault & Ritchie, 1980; Jenkins, 1987; Perdue, 1985). Interpersonal communication is frequently observed in vacation planning, and multiple sources typically are used to retrieve information (Capella & Greco, 1987; Myers & Moncrief, 1978; Nichols & Snepenger, 1988).

According to Snepenger et al. (1990), four major factors influence information search in the tourism context: (1) the composition of vacation groups; (2) the presence of family and friends at the destination; (3) prior visits to the destination; and (4) the degree of novelty associated with the destination. Family size and education are found to have a positive association with the use of destination-specific literature, while age (Gitelson & Crompton, 1983) and occupation (Woodside & Ronkainen, 1980) seem to correlate with use of travel agencies. Interpersonal communication dominates information acquisition for many travelers (Gitelson & Crompton, 1983), and importance of interpersonal sources seems to increase with age (Capella & Greco 1987). First-time visitors to a destination are observed to rely more on professional sources than seasoned visitors (Snepenger, Meged et al., 1990; Van Raaij, 1986). Novelty of vacation is reported to increase both the length of search and the number of information sources used (Engle et al., 1973; Snepenger, 1987).

In recent studies, Fodness and Murray (1998, 1999) observe that tourist search strategies correlate with the following task related factors: (1) locus of search; (2) timing of search; and (3) selection of information sources used. They also report that consumers relying heavily on previous experience or interpersonal communication utilize fewer sources on average than those primarily using external information sources.

There are few reported studies of how travelers combine conventional and electronic information sources while searching for product information. However, it could be deduced from expectations related to electronic markets that, on average, there should be less need to resort to other information sources when using electronic information sources than there is when using only conventional sources. Search strategies based on simultaneous information retrieval and product comparison (Whinston et al., 1997) imply a limited selection of information sources used. However, as we suggested previously, electronic information sources may not be mature enough for substituting conventional information sources with them. Rather, if they are used as complementary sources, the total number of information sources used should be higher for those consumers using the electronic sources.

Proposition 3

Consumers using electronic information sources utilize a higher number of information sources than those resorting only to non-electronic sources.

Methodology

In this chapter, we report the method and conduct of data collection. We also evaluate the data obtained by comparing key demographic variables with those from the 1999 census (Statistics Finland, 2000) and a recent Finnish Internet tracking survey (Taloustutkimus, 2001). Finally, we discuss the possible sources and effects of the biases detected in our data.

Data Collection

To collect the data, we conducted a mail survey for the period from December 2000 through January 2001. The subjects were obtained by drawing a random sample of 3,000 customers of a major Finnish Internet service provider (ISP) commanding nearly a 40% share of ISP markets in 2000. The sample frame was restricted to registered customers over 10 years of age. The logic of including minors into our sample was that children may have an active role in planning vacations, and, thus, they are eligible subjects for the study.

The initial sample was screened, and duplicate subjects (customers with multiple contracts with the ISP) were discarded. We first contacted the subjects by e-mail and provided them with both the URL of the HTML version of the questionnaire and another version to be printed. To motivate the subjects, we announced a lottery to be held among all subjects participating in the study. The prize was one gift certificate, FIM 3,000 of value (e 500), valid to a travel agent partnering with the ISP. Those subjects not responding to our query in the first round were contacted three weeks after sending the initial e-mail message. This time, we mailed them a paper version of the questionnaire, although they still had the option of using the HTML form.

After two attempts to contact the subjects, we had still failed to reach a number of them, which left us with a final sample size of 2,539 subjects. They responded fairly actively, and we tallied 871 completed questionnaires, 353 of which came through the HTML form and 518 by mail. Thus, the response rate was 34.4%, which we deem adequate.

Results

The prevailing theories concerning the Internet as a venue of electronic commerce imply that information retrieval is expected to be very easy on the Internet (Alba et al., 1997; Bakos, 1997). Consumers are expected to be able to easily initiate information retrieval, and it is even hypothesized that they may have little motivation to resort to other information sources, especially if travel costs are associated with using these sources (Lal & Sarvary, 2000). Thus, the first test of the assumed efficiency of the Internet as a product information source is to study how many of our subjects reportedly relied solely on Internet sources (i.e., World Wide Web and e-mail) while searching for travel services for their latest trip. We assigned our subjects to three groups by the information channels they used: only conventional channels, only electronic channels, and combined conventional and electronic channels. This categorization is shown in Table 1. According to our data, the group of consumers resorting to electronic

Table 1. Subjects grouped by information channels used

		Frequency	Percent	Valid Percent	Cumulative Percent
Valid	Conventional only	120	27.7	28.5	28.5
	Combined	297	68.6	70.5	99.0
	Electronic only	4	0.9	1.0	100
	Total	421	97.2	100	
Missing	System	12	2.8		
Total		433	100		

channels only is still very small; 4 subjects out of 421 reported using only electronic information sources.

Number of Information Sources Used

Even if most consumers seem to combine electronic information sources with conventional ones, we should test whether electronic sources lessen the need to resort to other sources. Thus, the number of additional information sources used to complement the electronic sources should be quite limited, if the electronic sources were as efficient as touted. The number of information sources used in our data was skewed to the left (Skewness = -0.422), and the distribution was quite flat (Kurtosis = -1.369). We deemed the departure from approximate normality to be substantial, and, thus, we used non-parametric tests to compare the two groups. The result of the Mann-Whitney U test (mean ranks: conventional sources only = 82.31, electronic and combined source = 262.31, z = -14.429, $p < 0.000$) suggest that the majority of subjects have used the electronic sources to complement conventional sources rather than in lieu of them.

One possible explanation for our finding is that consumers are still largely experimenting with electronic sources. Electronic travel service markets have emerged relatively recently, and it is possible that consumers have not yet accumulated enough experience to use these mechanisms to their full potential. On the other hand, some subjects reported that finding sellers in the first place is rather difficult on the WWW, if one does not know the precise addresses (URL) leading to electronic retailers or service providers. This notion demonstrates the need to resort to conventional sources, like advertisements, to locate sellers on the Internet.

Combinations of Information Sources

Based on the previous analyses, we conclude that the Internet-based consumer markets for travel services are not yet mature enough to lend themselves to rational and efficient information search endorsed by the hypotheses commonly related to the electronic consumer markets. The underlying value maximization principle commands higher levels of knowledge of the markets before efficient weighing of costs and benefits would be possible. Yet, for some reason, consumers seem to use electronic information sources while searching for travel services. To better understand the motivation for using electronic sources, we next explore which information sources are used in combination with the WWW and which are not.

- **Dependent Variable:** Use of the WWW as an information source during the search is the phenomenon we are trying to explain by relating it to other external information available to consumers, as well as variables suggested in the past literature as determinants of information channel choice. The use of the WWW is measured as a dichotomous variable (1 = used, 0 = not used).
- **Independent Variables:** The following information sources are included in our model as individual dichotomous variables: physical travel services retailer, phone service, e-mail, brochure, and advertisement. We also examined the demographic variables discussed in relation to data validation for inclusion in our model. Age of subject was the only demographic variable associated with use of electronic sources to search for travel services. This is in accordance with previous findings concerning the effect of age on adoption of Internet technology (Taloustutkimus, 2001).

We constructed a binary logistic regression model to study the interrelationships between use of the WWW as an information source and the independent variables. The model is of the following form:

$$\log it(\pi) = \alpha + \beta_1 AGE + \beta_2 OFF + \beta_3 PHO + \beta_4 EMA + \beta_5 BRO + \beta_6 ADS$$

, where

$AGE = subject's\ age$

$$OFF = \begin{cases} 1 & if\ visited\ a\ retailer \\ 0 & otherwise \end{cases}$$

$$PHO = \begin{cases} 1 & if\ used\ a\ phone\ service \\ 0 & otherwise \end{cases}$$

$$EMA = \begin{cases} 1 & if\ used\ e-mail \\ 0 & otherwise \end{cases}$$

$$BRO = \begin{cases} 1 & if\ used\ brochures \\ 0 & otherwise \end{cases}$$

$$ADS = \begin{cases} 1 & if\ used\ advertisements \\ 0 & otherwise \end{cases}$$

The estimated model is shown in Table 2. As can be seen, variables excluding use of brochures had an impact on use of the WWW as an information source. Age had an adverse effect, as could be expected. Young people typically are among the early adopters of technology, while eagerness to adopt tends to diminish as a function of age. Visits to physical retail outlets also had an adverse effect on the use of the WWW, or vice versa; those subjects using the WWW tended to avoid visiting physical retail outlets. However, use of the WWW and other information sources with no need for transportation were positively associated. Use of e-mail is, apparently, the source most frequently used in combination with the WWW. Users of the WWW also used advertisements more frequently, which we interpret to strengthen our previous assumption regarding the difficulty of locating sellers on the WWW. Finally, there was no discernible association between the use of brochures and the WWW. Using cut point of 0.67, our logistic regression model classified 83.4% of the subjects correctly, which is higher than the 67% of the maximum chance criterion or the 56% of the proportional chance criterion.

The regression model demonstrates that few consumers use the WWW as the sole source of information when searching for travel services. It is used, instead, to complement other sources, either conventional or electronic. Since use of the WWW seems to be associated with sources that limit the need to visit retailers, most notably e-mail, we next probe the decision criteria that are associated with use of electronic sources. Our previous analyses suggest that time constraints related to transaction may associate with the use of electronic sources as they are found to associate with direct channels in general (Eastlick & Feinberg, 1994). One of our subjects formulated the benefits of the WWW as well as its constraints as follows:

> *The Web based services are convenient for checking the availability of travel services. If there are no vacant seats, there will be no time lost in vain queries. Price information is easy to retrieve in the Web. E-mail contacts to travel agencies are almost mandatory.*

Table 2. Logistic regression model of combination of information sources

Variable	B	S.E.	Wald	df	Sig	R	Exp(B)
AGE	-0.222	0.062	12.813	1	0.000	-0.141	0.801
OFF	-1.025	0.366	7.859	1	0.005	-0.104	0.359
PHO	1.249	0.297	17.694	1	0.000	0,170	3.488
EMA	2.184	0.320	46.715	1	0.000	0.287	8.883
BRO[(a]	0.108	0.355	0.093	1	0.760	0.000	1.114
ADS	1.446	0.314	21.237	1	0,000	0.188	4.244
Constant	0.684	0.691	0.981	1	0.322		

Decision Criteria and Choice of Information Sources

To gain a better understanding of the motives that presently drive use of electronic media, we did pair-wise comparisons of the importance of related decision criteria. Our subjects had reported decision criteria they used for selecting the latest trip. The criteria were measured using Osgood scales.

The originally measured criteria were found to contain several anomalies. Criteria like price and quality showed strong positive correlation in our data. Since these criteria should be somewhat conflicting in most real-life situations, we decided to remedy the anomalies by scaling the variables as suggested by Hair et al. (1998). Thus, we subtracted the average criteria score from individual scores to obtain the relative importance of criteria. The converted variables were examined for any departures from normality and found that all variables conformed to approximate normality.

The group means were compared with independent samples t-tests. The results are presented in Table 3. As can be seen, there is little difference in revealed importance of price. Likewise, the subjects utilizing different information sources regarded quality as important in both groups. However, those subjects combining conventional and electronic information sources regarded speed of search as a more important criterion than those resorting to conventional information sources only. We conclude that consumers presently seem to use electronic information sources, on average, much for the same reasons as they use direct sales channels—to avoid the need to visit stores and, thus, to speed up the transaction.

The results suggest that consumers searching for travel services on the WWW utilize, on average, a relatively high number of information sources. Thus, the WWW does not presently appear to be the information source over all other sources, as touted in past literature. It is, rather, a complementary information source providing consumers with relatively easy access to electronic versions of travel catalogs and inventory information. It appears to be a valuable source, especially for those consumers who try to avoid transportation costs. Further-

Table 3. Decision criteria by information sources used

Criterion	Information source	n	Mean	Std. Dev.	t-value	df	p-value (two-sided)
Price	combined	278	0.106	1.730	0.177	370	0.858
	conventional	94	0.070	1.615			
Quality	combined	278	1.009	1.325	1.109	370	0.268
	conventional	94	0.836	1.267			
Speed of search	combined	278	-0.005	1.643	2.404	139.492 (a	0.018
	conventional	94	-0.547	1.966			

more, our results imply that most consumers extend their search to electronic markets to improve the search process, not the quality of their purchase.

Discussion

It appears that consumers do not presently conceive the WWW as an information source having a decisive advantage over all other sources. Rather, they use WWW-based services to supplement information they have found from other sources. Consumers seem to actively combine e-mail and WWW as information sources. One possible explanation is that e-mail makes it easier for consumers to more efficiently interact with sellers by sending requests for offers, for example. Thus, they rely more on the ability of travel service providers and travel agents to provide accurate information than on trying to search for information by themselves. This interpretation supports the hypothesis that electronic markets may increase simultaneous search strategies, as put forth by Whinston et al. (1997), yet human interaction is still part of the search. Rather than being a separate market, the Internet appears to offer new information channels to complement the conventional sources. These sources may alleviate some of the information search costs faced by consumers, while familiarizing with and comparing products, yet they do not enable consumers to engage in such efficient product comparison as is often expected in the electronic market literature.

It should be noted that consumers combining electronic and conventional sources in our sample used a relatively high number of information sources while searching for travel services. Thus, it is reasonable to expect that the WWW offers some unique benefits as an information source when compared to other sources. Some subjects commented that the WWW provides easy access to travel service inventories. Knowledge of the availability of services facilitates the negotiation process, since consumers are able to focus on available options only.

The typical negative comments dealt with locating travel-services-related WWW sites. It is true that, in this sense, the WWW is a rather demanding information source. Consumers may observe locations of physical agencies as they are in motion. Phone books provide exhaustive lists of sellers. While search engines and directory services may succeed in providing the same information, many consumers are better able to locate a seller in their physical neighborhood or in a phone book. The mental effort related to the first step of the search process may thwart those consumers who are least capable of locating sellers on the WWW. Evidently, many sellers also have recognized this problem. It is increasingly common that conventional advertisements include a URL for further information.

For business practitioners, the results of the study may be summarized as follows: (1) It pays off to provide product information through electronic channels, even if the majority of consumers still tend to abstain from buying through them. Most consumers with access to the WWW search it for product information, and the WWW is a cost-effective medium to reach those consumers; (2) The primary Web-related goal seems to be the desire to avoid moving about while searching for products; it may prove difficult to try to substitute electronic information channels with media, which do not meet this goal; (3) Electronic channels may not yet provide consumers with all the necessary information to make the purchase decision; hence, businesses should make sure that consumers are directed to other, complementing information sources when the electronic channels fall short of consumers' expectations; (4) Consumers should be provided with an option of retrieving product information through complementary information channels based on human-to-human interaction; e-mail and phone services should be particularly well suited for this, since they meet the previously mentioned goal consumers seem to attach to the use of electronic channels—to avoid the need to move about.

Limitations of the Study

The results of this study are exploratory and, as such, need further analysis for confirmation. More detailed empirical analyses by various travel service categories could provide a clearer picture of how the effects of Internet-based information sources affect the extension of consumer information search.

This study was limited to an investigation of how consumers use external information sources. Other factors should be imported to the study to obtain a better understanding of the role of Internet-based information sources in the consumer information search process. Use of external information sources should be related to the use of internal source (memory), external neutral sources (friends, relatives, consumer reports), and timing of the search process. The last dimension of search is particularly intriguing. It should prove beneficial to understand how time constraints affect the use of Internet-based information sources and to relate the use of these sources to both prepurchase and ongoing search.

Consumer information search is such a large target that any study is by necessity limited to providing only a glimpse of the whole complex. Even with its limitations, this study has attempted to address the most prominent dimensions of search effort and to succeed in providing some evidence in support of the assumptions that these new sources will affect consumer search. Since the perceived changes do not conform to some of the previous expectations, we believe that these results may help to refocus the academic dialogue concerning the effects of electronic markets on consumer search.

Topics for Future Research

This study has focused on the external consumer search process and the determinants of information source selection. However, the consumer search process is known to extend well beyond the scope of our framework. The locus of consumer search has received substantial attention in consumer behavior literature. Consumers are known to resort to an external information search typically after exhausting their internal information sources (i.e., long-term memory). Little is still known about how use of electronic media is affected by the use of internal and interpersonal information sources. Equally vague is our knowledge of how the electronic media affect the need for internal search and use of interpersonal information sources.

Use of electronic information sources with regard to the spatial dimension of information search is also largely unexplored. The timing and duration of information search are likely to affect the selection of the information sources used to retrieve information. Two categories of search—ongoing and prepurchase—often are used to denote the types of search processes in relation to their timing relative to the procurement decision.

The third major area of consumer search warranting future research with regard to electronic media is the search process itself. Consumers are implicitly expected to change their search strategies and to become more rational in the sense that rationality is commonly understood in the economic context. We still do not know whether and to what extent electronic markets will change consumer search strategies. Nevertheless, substantial deviation from the present behavior should result before the electronic markets can fulfill the high expectations placed on them.

References

Alba, J.J., et al. (1997). Interactive home shopping: Consumer, retailer and manufacturer incentives to participate in electronic marketplaces. *Journal of Marketing, 61*(3), 38-53.

Bakos, J.Y. (1991). A strategic analysis of electronic marketplaces. *MIS Quarterly, 15*(3), 295-311.

Bakos, J.Y. (1997). Reducing buyer search costs: Implications for electronic marketplaces. *Management Science, 43*(12), 1676-1692.

Bakos, J.Y. (1998). The emerging role of electronic marketplaces on the Internet. *Communications of the ACM, 41*(8), 35-42.

Balasubramanian, S. (1998). Mail versus mall: A strategic analysis of competition between direct marketers and conventional retailers. *Marketing Science, 17*(3), 181-195.

Bettman, J.R., Johnson, E.J., & Payne, J.W. (1990). Consumer decision making. In T.S. Robertson, & H.H. Kassarjian (Eds.), *Handbook of consumer behavior* (pp. 50-84). Englewood Cliffs, NJ: Prentice-Hall.

Bettman, J.R., Luce, M. F., & Payne, J. W. (1998). Constructive consumer choice processes. *Journal of Consumer Research, 25*(3), 187-217.

Brynjolfsson, E., &. Smith, M. D. (1999). The great equalizer? The role of shopbots in electronic markets. *WISE*, Charlotte, North Carolina, December 11-12.

Brynjolfsson, E., & Smith, M. D. (2000). Frictionless commerce? A comparison of Internet and conventional retailers. *Management Science, 46*(4), 563-586.

Bucklin, L.P. (1969). Consumer search, role enactment and market efficiency. *Journal of Business, 42*, 416-438.

Capella, L.M., & Greco, A.J. (1987). Information sources of elderly for vacation decisions. *Annals of Tourism Research, 14*, 148-151.

Celsi, R.L., & Olson, J.C. (1988). The role of involvement in attention and comprehension processes. *Journal of Consumer Research, 15*, 210-224.

Claxton, J.D., Fry, J.N., & Portis, B.A. (1974). A taxonomy of prepurchase information gathering patterns. *Journal of Consumer Research, 1*, 35-42.

Clemons, E. K., Hann, I.-H., & Hitt, L.M. (1999). *The nature of competition in electronic markets: An empirical investigation of online travel agent offerings* [working paper]. Philadelphia, PA: Wharton School of the University of Pennsylvania.

Dommermuth, W.P. (1965). The shopping matrix and marketing strategy. *Journal of Marketing Research, 2*, 128-132.

Eastlick, M.A., & Feinberg, R.A. (1994). Gender differences in mail-catalog patronage motives. *Journal of Direct Marketing, 8*(2), 37-44.

Engle, J., Kollat, D., & Blackwell, R. (1973). *Consumer behavior*. Hinsdale, IL: Dryden Press.

Filiatrault, P., & Ritchie, J.R.B. (1980). Joint purchasing decisions: A comparison of influence structure in family and couple decision making units. *Journal of Consumer Research, 7*(2), 131-140.

Fodness, D., & Murray, B. (1998). A typology of tourist information search strategies. *Journal of Travel Research, 37*(2), 108-119.

Fodness, D., & Murray, B. (1999). A model of tourist information search behavior. *Journal of Travel Research, 37*(3), 220-230.

Gitelson, R.J., & Crompton, J.L. (1983). The planning horizons and sources of information used by pleasure vacationers. *Journal of Travel Research, 21*, 2-7.

Hair Jr., J.F., Anderson, R.E., Tatham, R.L., & Black, W.C. (1998). *Multivariate data analysis*. Upper Saddle River, NJ: Prentice-Hall.

Haubl, G., & Trifts, V. (2000). Consumer decision making in online shopping environments: The effects of interactive decision aids. *Marketing Science, 19*(1), 4-21.

Hempel, D.J. (1969). Search behavior and information utilization in the home buying process. In P.R. McDonald (Ed.), *Marketing involvement in society and the economy* (pp. 241-249). Chicago: American Marketing Association.

Jenkins, R.L. (1987). Family vacation decision-making. *Journal of Travel Research, 16*, 2-7.

Kahneman, D., & Tversky, A. (1979). Prospect theory: An analysis of decision under risk. *Econometrica, 47*(2), 263-291.

Katona, G., & Mueller, E.A. (1955). A study of purchase decisions. In L.H. Clark (Ed.), *Consumer behavior: The dynamics of consumer reaction* (pp. 30-87). New York: University Press.

Kiel, G.C., & Layton, R.A. (1981). Dimensions of consumer information seeking behavior. *Journal of Marketing Research, 18*(2), 233-239.

Korhonen, P., Saarinen T., & Öörni A. (2000). *Tuotevertailu sähköisillä markkinoilla*. Helsinki: LTT-Tutkimus Oy.

Krugman, H. (1965). The impact of television advertising: Learning without involvement. *Public Opinion Quarterly, 29*(3), 349-356.

Lal, R., & Sarvary, M. (2000). When and how is the Internet likely to decrease price competition? *Marketing Science, 18*(4), 485-503.

Lee, H.G. (1998). Do electronic marketplaces lower the price of goods? *Communications of the ACM, 41*(1), 73-80.

Midgley, D.F. (1983). Patterns of interpersonal information seeking for the purchase of a symbolic product. *Journal of Marketing Research, 20*(1), 74-83.

Myers, P.B., &. Moncrief, L.W. (1978). Differential leisure travel decision-making between spouses. *Annals of Tourism Research, 5*, 157-165.

Newman, J.W. (1977). Consumer external search: Amount and determinants. In A.G. Woodside, J.N. Sheth, & P.D. Bennett (Eds.), *Consumer and industrial buying behavior* (pp. 79-94). New York: North-Holland.

Newman, J.W., &. Staelin, R. (1972). Prepurchase information seeking for new cars and major household appliances. *Journal of Marketing Research, 9*, 249-257.

Nichols, C., & Snepenger, D. (1988). Family decision making and tourism behavior and attitudes. *Journal of Travel Research, 26*, 2-6.

Perdue, R.R. (1985). Segmenting state travel information inquirers by timing of the destination decision and previous experience. *Journal of Travel Research, 23*(3), 6-11.

Rayport, J.F., & Sviokla, J.J. (1994). Managing in the marketspace. *Harvard Business Review, 72*(6), 141-150.

Schaninger, C.M., & Sciglimpaglia, D. (1981). The influence of cognitive personality traits and demographics on consumer information acquisition. *Journal of Consumer Research, 8*, 208-216.

Schmidt, J.B., & Spreng, R.A. (1996). A proposed model of external consumer information search. *Academy of Marketing Science Journal, 24*(3), 246-256.

Simon, H. (1955). A behavioral model of rational choice. *Quarterly Journal of Economics, 69,* 99-118.

Simon, H.A. (1957). *Models of man.* New York: Wiley.

Simon, H.A. (1982). *Models of bounded rationality*. Cambridge, MA: MIT Press.

Simon, H.A. (1987). Rationality in psychology and economics. In R.M. Horgarth, & M.W. Reder (Eds.), *Rational choice, the contrast between economics and psychology* (pp. 25-40). Chicago: The University of Chicago Press.

Slama, M.E., & Tashchian, A. (1985). Selected socioeconomic and demographic characteristics associated with purchasing involvement. *Journal of Marketing, 49*(1), 72-82.

Snepenger, D., Meged, K., Snelling, M., & Worrall, K. (1990). Information search strategies by destination-naive tourists. *Journal of Travel Research, 29*(1), 13-16.

Snepenger, D., & Milner, L. (1990). Demographic and situational correlates of business travel. *Journal of Travel Research, 28*(4), 27-32.

Snepenger, D., & Snepenger, M. (1993). Information search by pleasure travelers. In M.A. Kahn, M.D. Olsen, & T. Var (Eds.), *Encyclopedia of hospitality and tourism* (pp. 830-835). New York: Van Nostrand Reinhold.

Snepenger, D.J. (1987). Segmenting the vacation market by novelty-seeking role. *Journal of Travel Research, 26*(2), 8-14.

Statistics Finland. (2001). Finland in figures. *Statistics Finland.*

Taloustutkimus. (2001). *The Finnish Internet tracking*. Taloustutkimus.

Udell, J.C. (1966). Prepurchase behavior of buyers of small electrical appliances. *Journal of Marketing, 30*, 50-52.

Van Raaij, W.F. (1986). Consumer research on tourism: Mental and behavioral constructs. *Annals of Tourism Research, 13*, 1-9.

Vulkan, N. (1999). Economic implications of agent technology and e-commerce. *The Economic Journal, 109*(453), F67-F90.

Whinston, A.B., Stahl, D.O., & Choi, S.-Y. (1997). *The economics of electronic commerce*. Indianapolis, IN: Macmillan Technical Publishing.

Woodside, A.G., & Ronkainen, I.A. (1980). Vacation planning segments: Self-planning vs. users of motor club and travel agents. *Annals of Tourism Research, 7*, 385-393.

Chapter V

Consumers on the Road from E-Communication to E-Shopping

Kari Elkelä, Finland Post Corporation, Finland

Tuomas Kokkonen, Finland Post Corporation, Finland

Heikki Nikali, Finland Post Corporation, Finland

Abstract

This chapter deals with the effect of new forms of consumer communication on shopping behavior. New communication technologies are in a constant state of development, and much debate is taking place concerning the multi-channel communication society. However, in reality, people are continually switching from one channel to another. In this regard, what is important is from whose point of view the system eventually will be controlled. It also can be argued that the opinions and needs of ordinary people have not necessarily received the strongest focus in developing multi-channel technology. This chapter is divided into three main sections. First, the development in message volumes and new technologies within the past decade is analyzed. Second, consumer preferences concerning the means of communication when receiving marketing information are analyzed.

Third, the evolution and current state of B2C e-commerce and changes in consumers' purchasing habits are discussed. Finally, a future vision of consumer behavior in a multi-channel world is presented.

Introduction

In this chapter, we deal with the effect of new forms of consumer communication on purchase behavior. The subject of the examination is the development of communication means and its wider consequences (e.g., receiving advertising, paying invoices, and consumers' purchasing behaviors). The rapid development of information technology that began in the 1990s has had a profound influence on the ordinary consumer's work day. The Internet and, in particular, the mobile phone have become primary tools in communication between people, and their roles also have also increased considerably in consumer marketing. Trade taking place through the Internet has increased steadily, and this development seems to be accelerating. Indeed, it seems that there is a profound digital revolution in progress that affects consumers' lives and behaviors in several areas. Finland, on account of its advanced technology and people's positive attitudes toward technology, is participating in the first wave of this development. However, consumer preferences tend to change slowly and may require active guidance and realized experiences of utility. The direction and speed of development in other advanced countries also can be estimated from the changes that are now taking place in Finland.

Historical Development of Communication

If we consider the evolution of targeted communication over the past 20 years, we will notice many interesting stages. During the 1980s, the letter experienced a period of intense growth in use. Although letter services were rivaled by the fixed telephone, the increase in communication sufficed for both forms. Furthermore, at the same time, the prevalence of electronic communication generated new forms of letter messaging (Nikali, 1998).

A new competitor to the letter arrived in the 1980s in the form of the telefax, which enabled the printed message to be sent in real time through an electronic channel. This brought about two significant changes in the evolution of the letter:

the generating effect of electronic communication on letter messaging gradually disappeared, and letter volume growth diminished (Nikali, 1998).

In the middle of the 1990s, e-mail overtook telefax and pushed its volume down. E-mail had many new properties. For example, messages transmitted through e-mail were easier to reprocess. E-mail brought text-targeted communication, an entirely new form of interaction, almost in real time. It has challenged not only the paper letter but also the telephone in efficiency.

Developments in Finland

The penetrations of different types of communication methods in Finland are quite high, and, in many respects, are the highest in the world. During the last five years especially, mobile phones and e-mail have become commonplace. At the same time as the distribution of the mobile phone has increased, the use of the fixed phone network and the penetration of the fixed phone have fallen fast. Presently, 45% of Finnish households are without fixed telephones, when five years ago they were in almost every home (Statistics Finland, 2005). In May 2004, 94% of Finnish households already owned at least one mobile phone.

Comparison to International Developments

International penetration comparison of different communication means is shown in Table 1. It can be seen from this that Finland is clearly more advanced than EU countries on average.[1] The comparison shows a different course of development in Europe and North America. For the mobile phone, the United States and Canada fall behind many EU countries, but they are more advanced in Internet use. In particular, broadband access in the United States and Canada are ahead of Europe. The speed of change is described by the fact that in 2001, the number of mobile subscriptions in EU countries per 100 inhabitants was 74,

Table 1. Penetrations of communication means in Finland and internationally at the beginning of the 2000s

Communication Mean	**Penetration (percent)**		
	Finland	**EU**	**United States, Canada**
Fixed phone per 100 inhabitants, 2001	67	59	68
Mobile phone per 100 inhabitants, 2001	80	74	44
Internet penetration in companies, 2000	91	68	71
Internet penetration in households, 2001	39	27	44
Share of Internet users, 2002	58	39	62
Broadband access per 100 inhabitants, 2002	3,3	2,3	6,1

Source: Statistics Finland (2003a) and Statistics Finland (2003b).

whereas five years earlier it was only nine. The second example of rapid technological development is the fact that the number of broadband connections in Finland has tripled during the last two years (Statistics Finland, 2004a).

On the basis of Table 1, it can be said that the digital divide of communication is not a central problem in Finland when examined from the point of view of the distribution of technology. The same result also has been reached in a study of the expansion of the information society in Finland (Statistics Finland, 2003a). The change that has taken place in Finnish consumers' communication behavior in recent years has not been concerned merely with the distribution of devices, but rather that the new communication technology has spread evenly into nearly all age groups and social classes. Also, the uses of communication means are similar, irrespective of the user's age or economic livelihood. Furthermore, there are now significantly fewer regional differences.

When the volume of communication was measured by the number of sent messages, it took only seven years to double in Finland, according to the latest figures (Statistics Finland, 2004c; Statistics of Finland Post Corporation). An especially favorable economic growth affected this quick development. When economic growth began to slow down at the turn of the last decade, it was immediately reflected in the development of communication. During the past three years, growth has almost come to a standstill.

In the period during which the total amount of communication has doubled, the changes among different communication means have been much more dramatic than the development in total communication volume. The situation is presented in Figure 1. The biggest change has taken place between fixed and mobile

Figure 1. Volume of messages in Finland according to type of communication means, 1995–2003

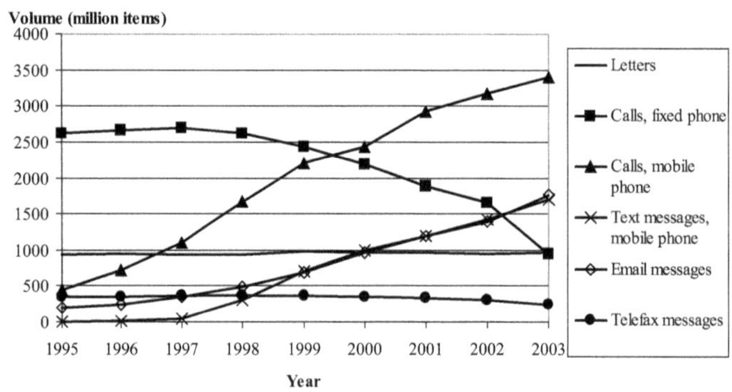

Source: Nikali – Heino (2003) and Statistics Finland (2003b)

phones. As late as the middle of the 1990s, fixed phones dominated the Finnish market; now, the focus is decidedly on the mobile phone. An especially dramatic fall has taken place in the volume of fixed phone calls in 2003. Text messages and e-mail have developed equally; text messages have risen from zero and e-mail from nearly zero to very significant positions. In this major change, letter services have retained their position quite well, even though this trend has been stable during the last years.

The Importance of Pricing

One of the most important factors in this development has been the price of services. In Finland, the real prices of domestic fixed phone calls rose from 1995-2004 by about 20%, while, at the same time, the real prices of mobile phone services halved (Ministry of Transport and Communications, 2004). Correspondingly, the real prices of letter services fell by about 3%. However, the price is not the most important factor in this development. Instead, the functionality and reliability of communication services are the most important factors in the choice of different communication means. One of the reasons is because the price development of communication services in Finland has fallen so as to diminish its effect. Table 2 shows the five most significant factors affecting the choice between different communication services in order of importance.

Furthermore, the attitudes of the Finns toward the new technology can be described well by the amount of money that has been spent on it. Since 1985, the real consumption expenditure of Finns has risen by 33% up to the year 2002 (Statistics Finland, 2003c). However, the real expenditure for communication services has tripled. The amount of money that has been used simultaneously for clothing has become smaller, and real expenditure on healthcare has increased only by a third. The growth of money that has been used for communication purposes has been directed only at telecommunications. Meanwhile, the amount of money that has been used for postal services has stayed at its present level since 1985. The money that has been used for telephone equipment has multiplied tenfold simultaneously, and the expenditure directed at telecommunication services has tripled.

Table 2. Five most significant factors affecting communications choice

Organizations	Households
1. Reliable Arrival of Message	1. Reliable Arrival of Message
2. Data Security	2. Data Security
3. Ease of Use	3. Price of Service
4. Speed of Communication	4. Ease of Use
5. Price of Service	5. Speed of Communication

If the prices of electronic communication services are compared internationally, Finland is one of the cheapest countries (Statistics Finland, 2003b). Finland is clearly below the EU and OECD countries' averages. However, the share of telecommunication in GDP is higher in Finland than in the other control countries. This shows that Finns also use more electronic communication services than consumers in the other countries.

Trends

Considering the future of communication in developed countries, some clear trends can be seen:

- The telefax is nearing the end of its life cycle.
- Spoken telephone calls will be completely transferred to mobile networks within a few years.
- The fixed telephone network will be used mainly for broadband connections.
- Contemporary SMS markets are already mature.
- The number of text messages is bound to rise. The increase, though, will take place in SMS-based services and not in personal communication.
- The number of e-mails will increase, and the quality will improve, as reflected in better security and junk e-mail filtering.
- The volume of letters will decrease, the major factor being the rate at which invoices will be sent in electronic form.

For the time being, the increase in electronic invoicing has almost come to a standstill, and the estimates clearly have been overly optimistic. The technology is ready, and companies are willing to adopt electronic invoicing because the cost savings are unquestionable. But the consumers are not yet ready for the change (Elkelä, 2004). In order to move on, companies have to offer consumers some incentive or use coercive methods. The vision of one network using a totally integrated technology cannot be regarded as a realistic option. Instead, different digital network solutions used simultaneously, or the so-called hybrid network, will be the prevailing trend. Although the technology necessary for this development already exists, it will take years to implement the change. This change requires large national investments and international agreements upon the introduction of common standards and rules.

Consumer Reception of New Marketing Channels

An important precondition for the development of e-shopping is the increasing use of e-messaging. During the last two years, three interesting studies have been published on consumer reception channel preferences in marketing. In an American study (Pitney Bowes, 2003), recipients were asked whether they preferred to receive their new product announcements via letter mail or e-mail. The letter was preferred by approximately 75% and e-mail by 20% of Americans.

It is most surprising that the popularity of letter mail and e-mail remained at the same level from 1999 to 2003, even though e-mail penetration has almost doubled over the same period. Also, in a Finnish study, it was shown that the percentage of letter receivers who would prefer electronic mail has stayed at a level of 5-6% from 1999 to 2003, in spite of the fact that e-mail penetration has almost doubled in Finland (Nikali & Heino, 2003). It seems that new e-mail users are less willing to use e-mail compared to experienced users, which means that substitution has been slower than anticipated.

Marketing Channel Preferences in Finland

In order to get a sense of the receptiveness of Finnish consumers to new channels of communication, Finland Post Corp. conducted research into the marketing channel preferences of consumers in Finland, focusing on four direct marketing channels: letter, phone call, text message (SMS), and e-mail (Elkelä, 2003). Typical marketing situations studied were those, for example, in which a consumer gets three holiday trip offers in a week, a new periodical offer, or a loyalty customer offer once a month. In this study, respondents were asked whether they accept channel use or disfavor it. Special attention was paid to examining which channel they not only accept, but also prefer (Figure 2).

Letter mail is overwhelmingly the most favored reception channel in every case. On average, it is preferred by 50-70% of respondents. Letter preference seems to increase if the marketing message is personal and important enough to the receiver. A good example of this kind of message is the loyalty card customer offer. Of Finns, 90% accept letter mail, while it is disfavored by only about 10%. This percentage remains roughly the same, regardless of the recipient's age, profession, total amount of communication, or Internet use. Advertising letters are not even disfavored by the most active e-mail users.

Figure 2. Preferred, accepted and disfavored reception channels for marketing offers in Finland (percent)

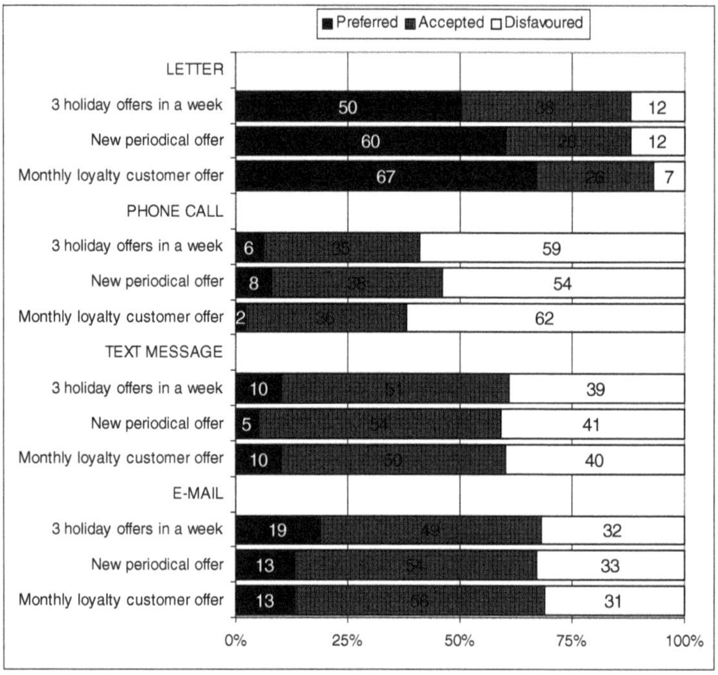

E-mail is preferred by 10-20% of respondents, although accepted by 70% of Finns. So, only 30% disfavor it. The letter and e-mail are the two channels chosen most. At the other end of the preference scale are phone calls. About 60% of respondents wish to avoid telemarketing, and less than 10% prefer it. Most consumers disfavor telemarketing, because in a typical situation, an unprepared customer feels irritated at being forced to face a well-prepared sales person putting him or her under pressure to make a quick decision.

Letters and e-mail messages are alike in many ways. Both offer marketing possibilities that do not interfere at the wrong time and can be read regardless of time and place. Further, unnecessary letters and e-mails can be both easily recognized and destroyed.

A letter is considered to have the advantage of greater security, and no equipment is required for its reception. The strongest points about e-mail are its speed, the possibility of avoiding paper waste, and lack of restrictions on place of reception. Nevertheless, the speed of e-mail is relative, because many holders of an e-mail address only read their home e-mail once a week or once a month. In such cases, letter communication for them is undoubtedly a faster reception channel.

Marketing Channel Preferences in Australia

A research conducted on Australian consumers (TNS Australia, 2003) showed quite similar preferences, in spite of the fact that this study also covered all unaddressed channels, including more than 10 marketing reception channels in all. The Australian study seems so far to be the only research in which respondents could choose from all possible reception channels. The research examines the effect of different customer message contents of which three are compared here. Because the Australian study allowed multiple answers, the total sum of revealed preferences can exceed 100%.

For loyalty program communication and information, the picture is quite similar to the Finnish case. The letter is preferred by about 70% of Australians and e-mail by about 10% (Figure 3).

Figure 3. Preferred marketing channels in Australia according to customer message contents (percent)

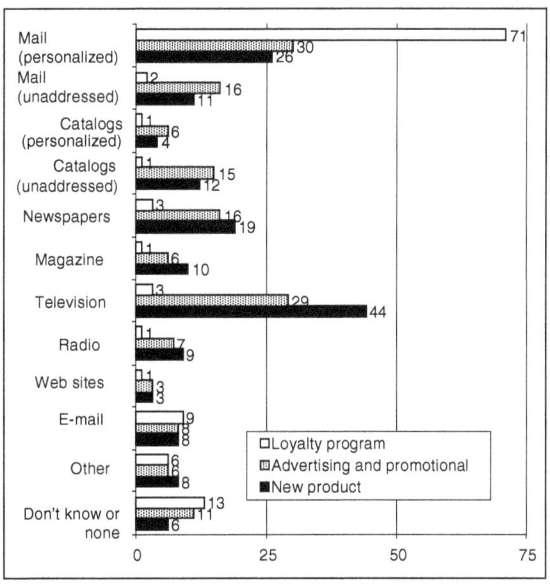

Figure 4. Preferred marketing channels in Australia according to industry whose products or services the customer uses (percent)

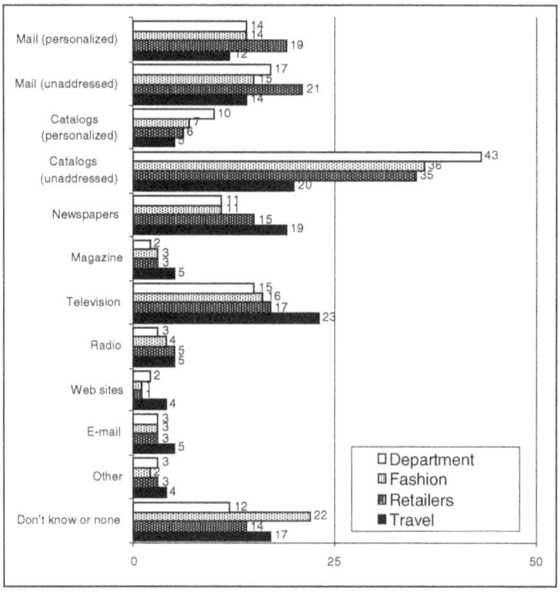

All other channels are of minor importance. The situation is different with new product announcements. Now, television becomes clearly the most preferred reception channel. Addressed mail is still the second. The next, in order of preference, are unaddressed mail, unaddressed catalogues, newspapers, and magazines, which are preferred by at least 10% of Australians. Regarding reception preferences for advertising and promotional material, addressed letter and television are again at the top. Unaddressed mail, catalogues, and newspapers are the second-most preferred group.

It is noticeable that for all these three message contents, only 8-9% of respondents prefer e-mail, and only 1-3% prefer the Internet for marketing message reception. The continuing unpopularity of these channels is a challenge for e-shopping. As long as consumers do not want to get information about products and services through these channels, they will not buy as much using these channels.

We can widen our view by looking at the reception preferences of different retail store and travel service customers (see Figure 4). For department stores, fashion stores, and retailers overall, unaddressed catalogues are the most preferred channel. Mail, newspapers, and television are also quite popular. Travel services, catalogues, newspapers, and television are just as popular. In these cases, e-mail is preferred by 3-5% of Australians and Web sites by 2-4%. There is no particular interest in receiving travel offers and information via e-mail, even if special offers for frequent flyers are desired by customers and have been successful in the airline business.

It seems that catalogues are considered the most convenient and entertaining marketing channel. Two other paper channels, mail and newspapers, are also highly preferred. Television is the only electronic channel preferred by over 10% of consumers. Mobile phones and their short text messages have been easy enough to adopt for consumers' daily communication with each other, but they are seen to be of little use in marketing and e-shopping. Only a few consumers seem to appreciate the possibilities of searching and real-time information offered via the Internet. The world seems to be full of e-mail addresses and broadband connections, but consumers are not so willing and able to use them effectively. Consumers prefer to keep their old customs, because effective use of new channels demands greater effort.

An obstacle to an explosive growth in e-marketing and e-shopping is that there are too few digital message pioneers who prefer to use mainly electronic messaging. Since, for example, e-mail reception preferences appear, for the present, to be at a standstill, an interesting question arises as to which factors probably would produce growth in the desire to accept e-mail marketing and at what stage this would take place.

Because consumers seem to want still more control over the reception of direct marketing, receiving messages through channels that are easy, agreeable, and not annoying will be emphasized in the future. The precondition for effective marketing will be based even more on knowing and respecting the channel preferences of recipients. In the future, senders will be required to introduce multi-channel solutions, which are able to target their marketing messages to any recipient's preferred reception channel.

To understand channel preferences and their meaning, further research and modeling of influencing mechanisms will, nevertheless, be required. For the time being, the popularity of marketing communication reception on paper has remained somewhat stable, even though there are many electronic alternatives to offer. A particularly important subject for future research would be the delineation of those factors that possibly trigger a change in the situation.

Figure 5. Finnish consumers who have made purchases over the Internet during the past three months (thousands of consumers)

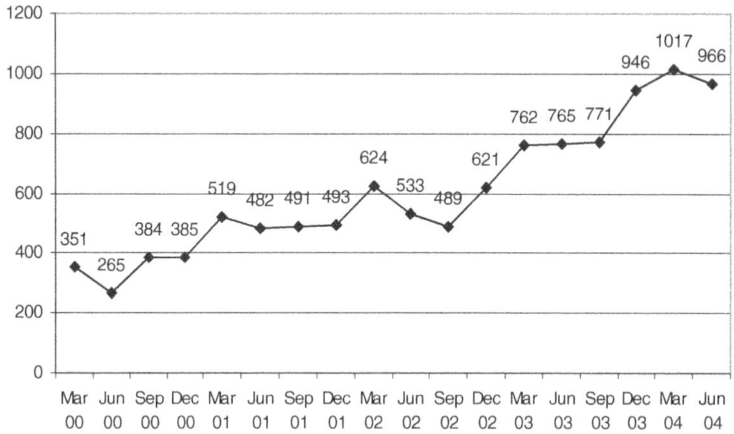

Will E-Commerce Change Shopping Habits? Assessing Potential Adoption

In addition to marketing channels, new technology may profoundly alter the consumer shopping habits, as well. At the turn of the 21st century, there were great expectations for Internet-based consumer business, but the outcome was less phenomenal, at least in the short run. The concept of consumer e-commerce was even hastily judged to be just another fad by several pessimistic observers. Now it seems, however, that the B2C e-business is actually alive and continuously growing at a steady rate, and will establish itself as an alternative shopping channel in the consumer business, after all. This insight can be illustrated by the fact that both the value of Internet purchases and the number of Internet shoppers has multiplied in the last few years. For example, the number of Finns who have made purchases on the Internet during the past three months exceeded one million in the first quarter of 2004, equaling practically 25% of all citizens between 15 and 79 years of age. This development can be seen in Figure 5 (TNS Gallup Oy, 2004).

Potential Barriers to E-Commerce

There are several existing barriers to e-commerce that may cause considerable delays to its adoption or even bar it altogether from some potential customer segments. In a recent special Eurobaromer report (European Opinion Research Group EEG, 2004), the lack of Internet connections was identified as the single most prominent obstacle for e-commerce in Europe, far more important than the next two obstacles; namely, lack of trust and lack of interest in shopping on the Internet. The availability of the Internet varied considerably between European countries, but, all in all, 57% of the citizens in the EU are still not connected to the Internet, which basically prevents them from buying goods online. This hindrance should be greatly alleviated by the decision of the European Union to actively promote the adoption of wideband Internet connections. Therefore, the prospects of B2C e-commerce seem quite bright today, but several questions remain: Will the growth of e-business turn into an exponential explosion, or will it stay at a conservative level in comparison to the total retail expenditure, after all? Moreover, what factors would make people prefer Internet-buying to the more traditional forms of shopping?

On several occasions, consumer e-commerce was supposed to be the most appropriate purchasing channel for those products that are intangible in nature and can be delivered to the customer in electronic forms via the Internet. Illustrative examples are computer programs and games and various tickets and travel reservations; growth in these areas has actually been faster than physical products. On the other hand, there have been deep doubts for e-commerce of those goods that consumers have become accustomed to touching, trying on, and testing in the retail stores. The most obvious examples of these products are naturally various types of apparel like shoes. In Finland, maybe a little surprisingly, recent research has shown that the second-most popular category of products bought over the Internet was actually clothes and shoes, which

Figure 6. Citizens in the EU-15 countries who have made purchases over the Internet (percent of population) in 2003

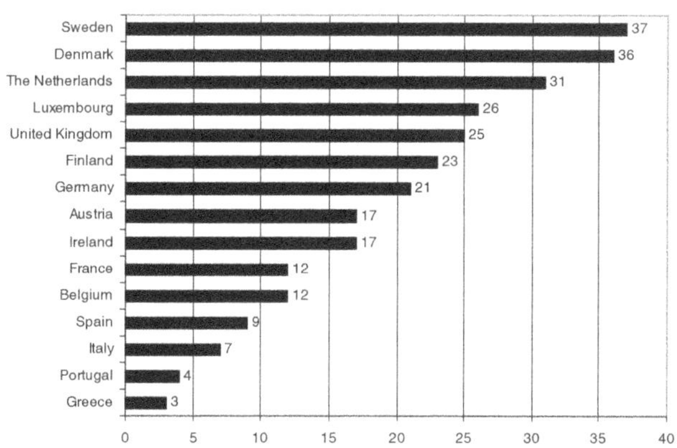

Figure 7. Percentage of product categories that have been bought by people shopping online in Finland (January-April 2004)

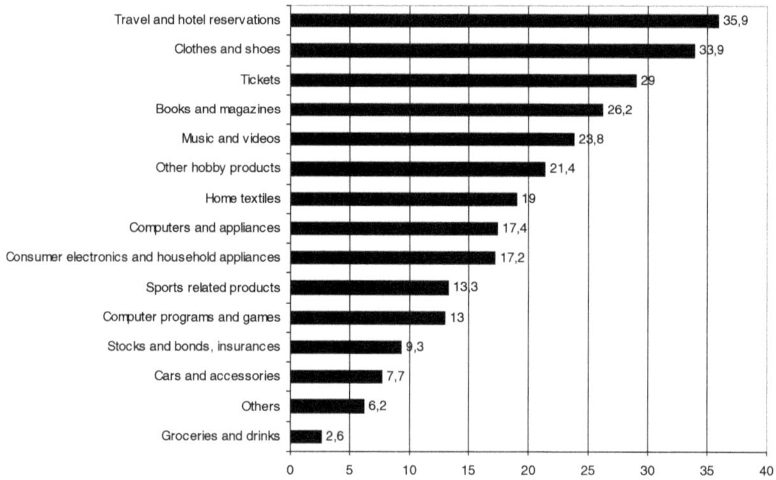

surpasses even relatively easily deliverable categories like books and CDs, as can be seen in Figure 7 (Statistics Finland, 2004d).

This finding seems to point to the conclusion that at least a considerable part of B2C e-commerce is actually carried out by the same persons that are customers in the traditional mail-order business (where the order is placed using, for example, a mail coupon or phone), because the same product categories are prominent in that business area, as well. This insight, however, is not fully agreed upon by the major companies engaging in both mail order and B2C e-business in Finland (e.g., Anttila and Hobby Hall). They claim that as many as two-thirds of their Internet customers are completely new, and only one-third consists of customers from the traditional mail-order business. On the other hand, as can be seen in Figure 7, there are several popular e-commerce product categories that have not been offered by mail-order companies, like various tickets and hotel bookings; so, in these areas, the Internet certainly provides a totally new buying channel.

Development of Alternative Channels for E-Commerce

The growth of e-commerce may not be limited solely to Web-based buying, as some other emerging technologies and ordering channels seem to point out. An especially interesting area is m-commerce (mobile commerce), which means making purchases with a mobile device like a cellular phone. The potential for m-business is estimated to be considerable because of the great number and penetration of mobile phones in developed countries today. At the time being, however, m-commerce is at a relatively low level and mostly experimental in nature because of the limitations in display resolution, rate of data transmission, and other central characteristics, which make mobile phones inferior to computer standards when it comes to making purchases with them. It can be expected, however, that as the capabilities of mobile devices improve in the near future, a noticeable share of purchases will be made using them, as people are becoming more and more accustomed to mobile phones as their one major mean of communication. It can be argued, although not yet empirically proved, that consumers will prefer to shop using the mode of communication that they use for receiving marketing material, for example. This means that the preferred channels and the consumer segments using them may be polarized further in the near future.

Digital television is yet another recent technological innovation that has been estimated to form a prominent ordering channel in consumer business as soon as the two-way data transmission is in wide use. Contemporary digital televisions are capable only of receiving data, but in the near future it will be possible for consumers to place an order with the TV remote control. This will open up completely new markets for distance sales, because television is a familiar, readily established, and widespread consumer interface with which people spend a considerable amount of their time daily. Advertisements with an instant ordering option very likely will increase impulse purchases and reach new customer segments who have no access to the Internet, the most obvious being elderly people.

Based on the discussion above, it is extremely difficult to estimate the growth rate of e-commerce and the effect that the new communication methods will have on consumer shopping behavior. Based on the available research, however, some notions can be articulated: First, although e-commerce seems to replace traditional mail-order business at least to some degree, there is a considerable number of consumers that have adopted distance sales (including both the traditional mail-order and more novel purchasing forms) only after the introduction of e-commerce and are, therefore, new customers in that area. That is why the share of distance sales can be expected to grow at a rate faster than that of traditional mail-order or common retail sales alone.

Second, in some product categories, the role of new communication modes and electronic purchasing will be considerable, whereas in some others the traditional ways of doing business prevail. A good example is buying food products on the Internet, which has been much less successful than expected a couple of years ago. On the other hand, the Internet is a great tool when searching and purchasing products that are rare or difficult to find in ordinary retail shops (e.g., various collectors' items, outdated spare parts, and local goods). Here, the Internet has opened completely new opportunities in global purchasing for ordinary consumers.

Third, as the capabilities of mobile phones and other portable devises improve, their impact in shaping consumer shopping habits probably will be of the same magnitude as with computers and the World Wide Web. Digital television probably will become another widely used interface, but this is unlikely to happen in the short-term.

Conclusion

During the last decade, a surprisingly simultaneous growth of two new modes of communication—e-mail and mobile phone—has occurred. The service options of the mobile phone later were extended through the introduction of the text message, which made it possible to send short written messages (SMS) and for them to be received anywhere. Although the length of a text message has been limited, the market has grown explosively over the past few years. At the same time, the price of a message has dropped quickly. Naturally, the uniqueness of the mobile phone is that it enables messaging, regardless of the location of the sender and the receiver.

The next big step in the evolutionary development of communication is an integrated means of communication and networks. However, this process has experienced a considerable setback over the past three years with the bubble of inflated market values bursting and a subsequent fallback into economic recession following. Nevertheless, it seems apparent that at the beginning of the next decade, different networks understand each other, and the present limitations on the transmitting of messages between different means of communication will, for the most part, disappear.

The possibilities for communicating will be considerably extended when communication is independent of the penetration of one type of device. The message sender will not even know through which device the message will be received. It is another matter, though, how consumers learn and desire to use the integrated type of communication systems and how effectively they can use them.

One of the critical aspects is connected with the term of community in communication. Targeted communication has been defined as a communication from a sender to specified receivers, and these all belong to the same community

(Nikali, 1998). Communication is even understood to be a communal commodity (i.e., the distribution of common impressions and experience in the community) (Viherä, 2002). An expedited message becomes actual and effective only when the receiver has deciphered the import of the message. If the receiver is unable to do this, he or she cannot interpret the message in the way intended by the sender.

It also can be argued that e-commerce will establish itself as an everyday ordering or purchasing channel, only after a critical mass of consumers have adopted some electronic medium (e.g., the Internet or, in the near future, a mobile device or digital TV) as their main communication technology. At that phase, the consumers will see the use of that particular channel as a completely natural and acceptable way of making purchases in addition to their other communication needs. Until then, e-commerce can be seen as just a modern-day version of the traditional mail-order business.

Development of Multi-Channel Customer Relationships

Our discussion concerning the effect of communication on shopping is only a starting point—with regard to the future use of multiple channels in customer relationships. Although the term *multi-channel* has been ambiguous, much debate has taken place concerning the nature of multi-channel customer relationship. When the definition is unclear, it will be understood in many different ways. The term *multi-channel* has been considered a generally acceptable concept. Nevertheless, rarely has it been discussed from whose point of view the multi-channel world will be developed. Does this take place based on the objectives of the companies, the consumers, or the service providers?

Usually, similar service channels, which are related to the care of the customer connections of companies, are referred to as the multi-channel customer relationship. The channels can be face-to-face meetings, connections that take place through electronic networks, or traditional functions based on physical transfer of documents and articles.

The definition of the multi-channel customer relationship naturally contains the idea of parallel use of channels. We could ask whether the recipient has a free choice of channel and a possibility to change his or her choice whenever he or she desires. It is more probable that companies prefer to use one or just a few channels based on their own requirements and cost savings.

In connection with the multi-channel customer relationship, the following terms would, indeed, have to be brought into use: a *democratic* and a *controlled multi-channel customer relationship*. A democratic multi-channel customer relationship refers to the situation in which the customer of the company chooses his or her service channel based on personal preferences. In the controlled relationship, the company sets the channels according to its own preferences and tries to direct customers to the most advantageous or most pleasant channel. The most efficient control methods are price and the quality of service.

One can ask justifiably if the controlled multi-channel customer relationship is multi-channel at all. The directing, indeed, is only a means to move from the model of one channel to another, to the channel model that produces more advantage for the companies. In the B2B sector, there usually is not a conflict in that sense, because the advantage is reached by both parties. In that case, the multi-channel model will become a genuine substitute. In the B2C sector, the channels are complements, but they also serve mainly as substitutes through the control.

References

Elkelä, K. (2003). *Paperilla vai sähköisesti? Osoitteellisen suoran halutut ja torjutut vastaanottokanavat. (On paper or electronically? Desired and undesired reception channels for direct marketing).* Helsinki: Finland Post Corp. Research Publications.

Elkelä, K. (2004). *Laskut ja kuluttaja. (Invoices and the consumer).* Helsinki: Finland Post Corp. Research Publications.

Elkelä, K., Jakosuo, K., & Nikali H. (2001a). *Kotitalouksien viestinvälitystarpeet 2001.* Helsinki: Finland Post Corp. Research Publications.

Elkelä, K., Jakosuo, K., & Nikali H. (2001b). *Organisaatioiden viestinvälitystarpeet 2001.* Helsinki: Finland Post Corp. Research Publications.

European Opinion Research Group EEG. (2004). *Issues relating to business and consumer e-commerce* [Eurobarometer special report 60.0/wave 201, 2004].

Insead. (2004). *eEurope 2005, A study of the degree of alignment of the new member states and the candidate countries.* Retrieved September 29, 2004, from *http://www.euractiv.com/ndbtext/infosoc/insead-sapeeurope2005 %20report.pdf*

Ministry of Transport and Communications. (2004). *Suomen telemaksujen hintataso 2003.* Helsinki: Research Publications.

Nikali, H. (1998). *The substitution of letter mail in targeted communication.* Helsinki: Finland Post Ltd., Quality and Business Development.

Nikali, H. (1999). The future of targeted communication in Finland. In M.A. Crew, & P.R. Kleindorfer (Eds.), *Emerging competition in postal and delivery services* (pp. 245-264). Boston: Kluwer Academic Publishers.

Nikali, H., & Heino A. (2003). *Kirjeliikenteen sisältötutkimus. Vuoden 2003 loppuraportti. (Content survey of letter communication).* Helsinki: Suomen Posti Oyj.

Pitney Bowes Inc. (2003). *Household mail preference study. Regular mail vs. electronic mail.* Stamford: Pitney Bowes Inc. Research Publications.

Statistics Finland. (2003a). *A great migration to the information society, patterns of ICT diffusion in Finland in 1996-2002.* Helsinki: Valopaino Oy.

Statistics Finland. (2003b). *Telecommunications in Finland 2002.* Helsinki: Multiprint.

Statistics Finland. (2003c). *Kotitalouksien kulutusmenojen muutossuunnat.* Helsinki: Yliopistopaino.

Statistics Finland. (2004a). *Consumer barometer study.* Helsinki: Statistic Finland Publications.

Statistics Finland. (2004b). *Suomalaisten viestintävalmiudet 2000-luvun vuorovaikutusyhteiskunnassa, Patterns of ICT diffusion in Finland in 1996-2002.* Helsinki: Valopaino Oy.

Statistics Finland. (2004c). *Telecommunications in Finland 2003.* Helsinki: Multiprint.

Statistics Finland. (2004d). *Tietoaika 8/2004.* Helsinki: Multiprint.

Statistics Finland. (2005). *Consumer barometer study.* Helsinki: Statistic Finland Publications.

TNS Australia. (2003). Reaching consumers in the information age. How Australian consumers prefer organisations to communicate with them (Commissioned by Australian Post).

TNS Gallup Oy. (2004). Web Commerce Trends.

Viherä, M.L. (2002). Viestintä edellyttää jäsennystä. In M. Mäkinen, K. Salminen, & M.-L. Viherä (Eds.), *Tietoyhteiskuntaa ymmärtämässä* (pp. 103-104). Helsinki: Nykypaino Oy.

Endnote

1 Denmark, Sweden, the Netherlands, United Kingdom, Finland, and Germany are the leaders in Europe on the basis of their ICT development and Internet usage (eEurope2005). In the study, an eEurope index was calculated. It is a composite of five key indicators: Internet, Modern Online Public Services, Dynamic eBusiness Environment, Secure Information Infrastructure, and Broadband. The study is a comparison among 28 European countries. It is essential for the leading countries to be near the top in each indicator. Correlation between eEurope index and GDP is very high. All the leading countries also are ranked in the highest bracket of GDP per capita.

Chapter VI

The Effects of Digital Marketing on Customer Relationships

Marko Merisavo, Helsinki School of Economics, Finland

Abstract

Viewing the use of digital channels in marketing from a customer relationship perspective offers several benefits to a marketer. Brand communication can be frequent and personalized, and different options for a dialogue exist. The benefits of dialogue include learning from and about customers, revealing their needs and interests, and being able to provide them with better and more personal service. From a customer perspective, the Internet has given power to get up-to-date information, to compare products and services more easily, and to get in touch with marketers. In a digital environment, customers increasingly use self-service and have more options to initiate action. A customer's perceived value of using digital channels and interacting with a marketer can come in several ways; it could be money, time, information, convenience, and so forth. This chapter examines how marketers can use digital channels to develop and strengthen customer relationships. In our framework, brand communication, service, personalization, and interactivity are suggested to be the key elements.

Introduction: Digital Marketing and Customer Relationships

According to Schultz (2002), multi-channel marketing is just a new term with old challenges; it appears to be 21st-century terminology for how a marketing organization makes its products and services available to customers and prospects and how the marketer determines the best choice of distribution systems and type of communication program to use. However, nowadays, the use of digital channels like the Internet, e-mail, mobile phones, and digital television in marketing has increased, and marketers need to understand how to use these channels for different purposes. For example, some marketers use them for acquiring new customers, while some focus on serving customers better. Digital marketing communication typically has been information about products, advertising on the Web, and direct marketing via e-mail and mobile. It seems like the channels and concepts of digital marketing have been used separately and operationally, while the strategic understanding and models of how and why to use different digital channels are still developing. At the same time, marketers have acknowledged the importance of cultivating customer relationships. Consequently, there is now increasing discussion about how to use digital marketing to develop customer relationships, an approach that has not been fully utilized yet.

To look at the use of digital channels in marketing from a customer relationship perspective offers several benefits to a marketer. Brand communication can be frequent and personalized, and different options for a dialogue exist. The benefits of dialogue include learning from and about customers, revealing their needs and interests, and being able to provide them with better and more personal service. From a customer perspective, the Internet has given power to get up-to-date information, to more easily compare products and services, and to get in touch with marketers. Furthermore, in a digital environment, customers increasingly use self-service and have more options to initiate action (e.g., give feedback, check account balances, request and personalize marketing communication). From the customer perspective, perceived value of using digital channels and interacting with a marketer can come in several ways. It could be money, time, information, convenience, pleasure, entertainment, assistance, social interactivity, prestige, or something else the customer appreciates. This chapter examines how marketers can use digital channels to develop and strengthen customer relationships. In our framework, brand communication, service, personalization, and interactivity are suggested to be the key elements.

Boosting Customer Relationships with Digital Marketing

Central Elements of Digital Marketing

We define *digital marketing* as marketing and interaction via digital channels—Internet, e-mail, mobile phones, digital TV, and so forth. Digital marketing includes communication (e.g., advertising, newsletters), service, and tools for customer-managed communication and interaction (e.g., FAQs, games, etc). Both marketer and customer can initiate interaction. Digital marketing as a term is more inclusive than e-marketing, because it encompasses marketing via digital channels independent of the medium.

Interactive marketing, one-to-one marketing, customer relationship management (CRM), brand management, and service management literature emphasize the importance of building and managing customer relationships, and each offers useful models to understand how customer relationships develop. The literature reveals certain concepts and trends that could be central elements in boosting customer relationships with digital marketing. First, brand communication is identified as an important way to create customer loyalty. Second, service is considered a central element of customer relationships, and many examples show that digital channels (i.e., the Internet) can be good tools for customer service. Third, personalization is expected to create value for customers. Fourth, interactivity is presented as one of the major advantages of the Internet and other new marketing channels. We think that brand communication and service are the key elements when boosting customer relationships via digital marketing, and elements of personalization and interactivity can be mixed into them to reinforce the effects (see Figure 1). We suggest that digital marketing can have a positive effect on customers' brand loyalty (i.e., purchases, attitudes, etc.).

Figure 1. Central elements in boosting customer relationships with digital marketing

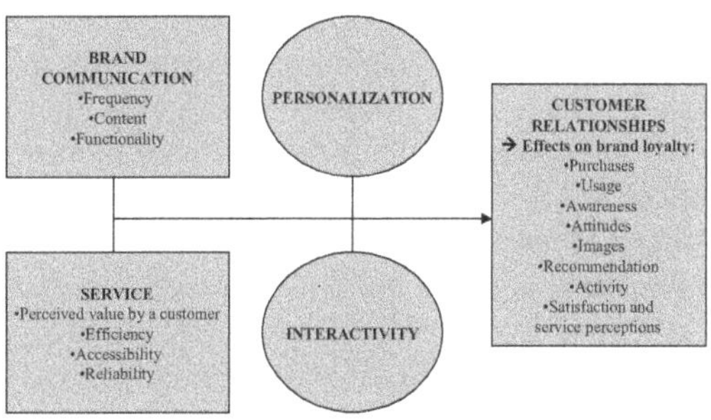

Three main factors in brand communication are expected to affect customer relationships: frequency (e.g., how many brand messages per week), content (e.g., promotional offers or information), and functionality (e.g., how easily customers find information on a Web site). Brand communication also can be personalized. For example, segmentation or customer profiles can be used to create different messages for different customers, sent via their preferred channels. In addition, brand contacts can be interactive, which means that customers can search for information, make inquiries, give feedback, and engage in various other activities with marketers or each other by playing games on a brand's Web site, chatting with other brand users, and so forth.

Service that is perceived as valuable by customers is expected to affect customer relationships positively. This value could be money, time, information, convenience, pleasure, entertainment, assistance, and so forth. Efficiency and accessibility of the service are also important, which could mean easy access to a Web site and the ability to find desired service options and information associated with them with minimal effort. Reliability of the service is essential, as well (e.g., the technical functioning and safety of an online bank). Service also can be personalized (i.e., by offering different service options for different customers and remembering their individual preferences of service situations). Various elements of interactivity (e.g., online chat with service personnel) can be combined with service.

In the model, brand communication and service are treated as separate elements, acknowledging that sometimes digital communication can be considered as a service, as Brondmo (2000) suggests for permission-based e-mail, while in other cases, it may be seen as a promotion. Brand communication and service do not always have to be personalized or interactive in order to have effects on customer relationships. Adding personalization and interactivity, however, can increase the positive effects of digital brand communication and service on customer relationships.

Brand Communication

Digital channels offer opportunities to keep up frequent brand communication with customers. Brand communication refers generally to all communication between the brand and customers. This complies with the view that the process of building brands and customer relationships is much more than traditional media advertising (Aaker & Joachimsthaler, 2000). Digital marketing can be used to create brand communication that customers perceive to be informative, entertaining, and interesting, and that building interactive brand contacts. Frequent communication has positive effects on customers' brand loyalty, and customer-brand relationships are deepened (Aaker, 1996; Kapferer, 1998).

As in the off-line environment, there is word-of-mouth communication in the digital environment, one visible form of that being a brand community. Brand communities are important for a marketer, because brand knowledge and experiences are quickly spread via those communities, and new opportunities for brand communication are opened. In brand communities, customers interact with each other and brands by communicating via different channels or face-to-face. In many cases, consumers and marketers jointly build brand communities. These communities may form around any brand, but it is more likely to form around brands with a strong image and a rich and lengthy history that threaten competition and are publicly consumed (Muniz & O'Guinn, 2001). Marketers are encouraged to create brand communities, because customers belonging to them are more loyal than others; they are less apt to switch brands and more motivated to provide feedback (McAlexander et al., 2002).

The digital environment offers great tools for community building. Virtual communities can have greater geographical scope and narrower focus than most off-line communities (Wind et al., 2002). Nevertheless, brand communities often mix the off-line and online environments, as community members often meet and discuss both on the Web and in the real world. The importance of community Web pages to consumers' day-to-day lives is demonstrated by their rich content, continuous update and modification, and active discussion. For example, Dynamiitti.com is a Finnish brand community site that connects the drivers and enthusiasts of five Japanese car brands (i.e., Honda, Mazda, Mitsubishi, Toyota, and Nissan). The site offers a discussion board with various topics for each brand, as well as areas for general discussion of driving and racing. Members share their views of new models, accessories, repair services, tests, and so forth. Generally, news spreads very quickly within this community. There also is a market forum, where community members can buy and sell cars and other items. This community converges virtual and physical communities, as active members regularly meet each other and join events in the bricks-and-mortar world.

Brand communities represent both opportunities and challenges for companies. There is an opportunity to build an online infrastructure where brand users freely share their brand knowledge, experiences, and feelings with each other. This information can be collected easily by the marketer, which gives an important inside view for the customers. Active community members can be considered opinion leaders who can be used to test new ideas and concepts and, hopefully, to spread positive word-of-mouth comments. There is also an opportunity to communicate directly to the brand users. For example, a representative of the marketer can be an active member of the brand community who joins everyday discussions among the other members. Also, ads can be placed for a sponsored brand community site. On the other hand, a company's ability to control the discussion of brand communities can be relatively limited or non-existing. Even on sponsored brand community sites, advertising must be kept at a minimum to keep members from being bombarded by too many commercial messages.

Service

The development of the Internet and other new channels has made it possible for firms to create new services and develop interactive and relationship-building contacts with their customers. To further emphasize the importance of online service, extensive survey data from more than 2,000 e-tail sites shows that price-sensitive customers may be the least loyal, whereas customer service support is the main factor that attracts repeat buying (Reibstein, 2002). Indeed, the most experienced and successful companies using the Internet are beginning to realize that the key to success is not based only on presence or low prices but, instead, on delivering electronic service quality.

The Nature of E-Services

Information service is that aspect of service in which information is the primary value exchanged between two parties (i.e., buyer and seller), and it is currently the highest growth area in service (Rust & Lemon, 2001). For example, the Internet is basically a network that allows the interchange of data and information, and many Internet services are quite intangible in their nature (e.g., search engines, news sites, e-magazines, online brokerage). Overall, digital channels offer many possibilities to provide service and go beyond the ordinary functions like access to goods and order fulfilment.

The digital environment is also changing the way customers interact with firms to create service outcomes. The importance of self-service is increasing, as customers more often use online banking and automated hotel checkouts, reserve flights over the Internet, track their packages, and so forth. That is, today's customers commonly interact with technology, not with the service firm employees, to create service outcomes. Customers also can actively create or shape the form of their services (i.e., personalize). For example, Yahoo! allows users to track their own stock portfolios, manage addresses and calendars, and follow news and information of interest to them in order to create their own service packages.

Valuable E-Services

Similar to the off-line world, marketers are now applying different techniques to create better service experiences in the digital environment. The focus of digital marketing is now shifting from creating transactions to creating service. In effect, satisfaction with e-service is most often driven by the fact that customers perceive it better than the alternative (i.e., interpersonal method of service delivery). Particular benefits include saved time, ease of use, and, for some customers, even avoidance of service personnel. In fact, many customers are very satisfied using e-services. A study sponsored by CRM vendor Kana found that 56% of consumers stated that interactions over the Web or via e-mail represent their most positive customer experience (CyberAtlas, 2002). Rapid response times to customer inquiries were recognized to be important for providing a positive customer experience. Similarly, in a study we conducted of service channels of a mobile operator's customers, we found that 50% of customers preferred the Internet or e-mail over the telephone, because they felt these channels were more efficient and flexible, offered better tools for documentation, allowed more control, were more convenient, and saved time (Pesonen, 2002).

Overall, the service perspective seems to be a very important and contemporary issue for marketers who use digital channels. In a survey among 33 leading Finnish companies from different parts of the value chain (i.e., marketers, media, etc.), managers were asked about their experiences and views on what kind of digital marketing works. They most commonly answered, "Marketing that creates the most added value to customers," and they clearly identified service to be one way to create that added value (Merisavo et al., 2002). The interviewed managers recognized the following e-services that are valuable to customers:

- Entertaining and useful online worlds (e.g., food- and cooking-related Web sites)
- Ability to check and change personal information
- Highly targeted offers or invitations
- Constructed environments for social interactivity (e.g., games, chats, etc.)
- Personalized services and communication in the form of own mobile phone model
- Real-time info on sports results
- Providing some exclusivity to customers via the use of digital channels (e.g., sneak previews, rumors, etc.)

In effect, these results match well with previous research. In a study conducted among 1,211 online customers, the following factors were found to positively affect loyalty toward an e-retailer (Srinivasan et al., 2002):

- **Customization:** Tailoring products, services, and transactional environments to individual customers.
- **Contact interactivity:** Availability and effectiveness of customer support tools and the degree to which two-way communication is facilitated.
- **Cultivation:** Providing relevant information and incentives to customers.
- **Care:** Attention to details in orders, resolving breakdowns, informing the status of orders.
- **Community:** The extent to which customers are provided with the opportunity and ability to share opinions among themselves through comment links, buying circles, and chat rooms provided by the e-retailer.
- **Choice:** The ability to offer a wide range of product categories and great variety of products.
- **Character:** Creative Web-site design creating an overall image or personality through the use of inputs such as text, style, graphics, colors, logos, and slogans.

Most of these antecedents of e-loyalty—7Cs, as Srinivasan and others call them—are service-oriented, which gives further evidence of the importance of e-service in building customer relationships. Also emphasizing the importance of service over other features when building customer loyalty via digital channels is the interesting evidence suggesting that sensory attributes, particularly visual cues, will influence choice to a lesser extent online than off-line (Degeratu et al., 2000). This result suggests that designing eye-catching Web sites is not the key to success; rather, it is service and content that drive purchases and build customer relationships online.

Personalization

One of the advantages of digital channels is the ability to identify individual customers and gather information from them and about them. Personalization is expected to be an element that can work to boost customer relationships by creating more personal, interesting, and relevant brand communication and better service. Therefore, in this section we focus on how personalization mediates the effects of brand communication and service on customer relationships.

What We Mean by Personalization

Personalization as a concept is rather vague. Different terms are used when talking about it. For instance, the following terms relate closely to personalization: customization, targeting, segmentation, profiling, and one-to-one marketing. It is beyond the scope of this chapter to explore the differences and relations of these terms. Instead, it is enough to acknowledge that personalization can mean different things to different people. Furthermore, it is important to see that personalization is not a one-shot action; rather, it is a process of interacting and learning between customers and marketers. It is helpful to divide personalization into three broad categories:

- Personalization executed by customers who fill their own preferences and state their interests; the company uses these profiles to address them individually (e.g., newsletters, Yahoo!'s customized Web pages).
- Personalization executed by a marketer, based on customers' behaviors or characteristics (e.g., targeted promotions based on buying behavior, special offer on a customer's birthday).
- Predictive group personalization (i.e., collaborative filtering) executed by a marketer (e.g., recommendations, offers, communication, etc.), based on the similar behavior, interests, or characteristics of other customers (e.g., book recommendations to anonymous browsers of Amazon.com).

On the Internet, many forms of personalization can be made automatic by systems and software. Similar to what was already presented, Parsaye (see Mohammed et al., 2002) has conceptualized personalization of Web sites as follows:

1. **Customization:** The system's ability to customize items by allowing individual users to set their own preferences.
2. **Individualization:** The system's ability to customize itself to the user, based on the user's exhibited behavior.
3. **Group Characterization:** The system's ability to customize itself to the user, based on the preferences of other users with similar interests.

To have positive results with personalization on customer relationships calls for customers to really appreciate personalization. There is some evidence that they do. For example, a study sponsored by CRM vendor Kana found that almost 75%

of respondents cited personalization as a major contributor to their most satisfying purchasing experience, online or off-line (CyberAtlas, 2002). The study was based on a cross-section of consumers who have made purchases in both online and off-line environments. According to the study, personalization capabilities, including self-service, personalized voice or e-mail interactions, the ability to track purchases and requests, and knowledgeable customer service representatives who are educated in the customer's history with a business, contributed to 73% of the most positive customer experiences. More than one-third of the respondents noted that lack of personalized customer care contributed to their least satisfying customer experience.

Personalizing Brand Communication

One clear object of personalization in brand communication is the content of messages. Content in digital media seems to be very important, as it is argued that consumers tend to be more critical when evaluating content on the Web than in print media (Gallagher et al., 2001).

The personalization possibilities of digital channels can be used to produce more relevant messages to customers, even according to their individual wishes. Nevertheless, it seems that personalization has focused too much on selling and transactions. Although tailored offers and promotions can be very valuable for customers, they also welcome richer content. That is, customers may appreciate information on new products, tips on usage, news related to their lifestyle and spare-time activities, invitations to events, and so forth.

Indeed, a survey of e-mail marketing of a cosmetics brand showed that loyal consumers appreciate regular communication and various other information content from the brand, not just offers (Merisavo, 2001; Merisavo & Raulas, 2004). In more detail, consumers found the following messages especially useful: special sales offers (90% of all respondents), information about new products (89%), contests (68%), news about beauty in general (68%), information about events (43%), links to Internet pages (43%), and information about international makeup trends (41%). Moreover, the desired content links to communication frequency; those consumers who appreciate regular contact want to hear on all of these items from the marketer. On the other hand, respondents who don't appreciate regular communication may find special sales offers and competitions especially useful but may not appreciate other content of communication as much. That is, it may be worthwhile for a marketer to consider personalizing brand communication, both in its frequency and content.

Personalizing Service

There can be different levels of personalized services. Sterne (2000) recognizes the following personalization levels:

- Recognition (greetings, name)
- Making recommendations (based on collaborative filtering)
- Making customer service recommendations (solutions to problems)
- Profile access; password required (customer-modifiable profiles, dynamic content)
- Explicit vs. implicit information (what customers tell and what can be objectively perceived of them)
- Anticipation engines (proactive personalization based on sophisticated data analysis)

This list shows that there are many ways to personalize service in interactive media, and the requirements are different. Some personalization options require extensive information from the customers, but others also work on anonymous users. In addition, the requirements for hardware, software, and competences are different for the marketer. There is not one way to create personalized services that work; instead, they need to be tailored for each business case. A Lands' End case is a great example and success story of how personalized service works in an online business. Lands' End is a direct merchant of traditionally styled, casual clothing for men, women, and children, as well as soft luggage and products for the home. Lands' End embraced the Internet early on, opening its e-commerce business in July 1995, featuring only 100 products. Internet sales now account for approximately 20% of its overall sales, and they have become the leading online seller of apparel in the world. The site has grown consistently, offering every Lands' End product, featuring personalized shopping accounts and hosting 38 million visits in fiscal year 2000. Landsend.com has extended Lands' End's one-on-one relationship with customers via personalized shopping aids such as Lands' End Live (talk to a customer service agent while shopping online), Your Personal Model (build a wardrobe based on individual shape and lifestyle), and Oxford Express (get help selecting dress shirts).

Interactivity

The Nature of Interactivity and Its Effects on Loyalty

Digital environment offers opportunities to turn communication from one-way to interactive. Interactivity may consist of different elements, depending on the channel and media being used. For example, on Web sites, interactivity can be in the form of customer support tools (e.g., information, search processes, etc.) and two-way communication (e.g., contact forms, chat, etc.), as well as games and entertainment. Customers are engaged in a search or dialogue with firms and brands, where information and actions flow both ways. The firm is not always the party that starts the interaction or dialogue; instead, in many cases, customers act as initiators. It would be beneficial to explore whether interactivity affects a customer and brand loyalty. Research suggested that interactivity has an impact on customer loyalty, at least for two reasons:

- Interactivity enables a search process that quickly can locate a desired product or service, thereby replacing dependence on detailed customer memory (Alba et al., 1997).
- Interactivity increases the amount of information that can be presented to a customer (Deighton, 1996; Watson et al., 1998).

Stewart and Pavlou (2002) stress the different nature of measurement in interactive media. "Research that treats marketing communication as an independent variable, useful for predicting consumer response, ignores the reciprocal influence that consumer response has on subsequent communication" (p. 381). That is, any response to marketing communication, including that of simply attending, may be contingent on other factors. Perhaps due to this complexity, the psychology behind the development of customer loyalty on the basis of interactive marketing methods is not well understood.

Kwak et al. (2002) found that product information requests over the Web are positively related to online purchases. When customers can search easily, find desired services or support, and filter out the information they need, they see this as a convenience and better service. In effect, companies are increasingly offering customers interactive tools to make better decisions about financial issues, purchases, health care, and other complex issues. The same is true in B2B sector; for example, eRoom provides collaboration tools in the form of a digital workplace to help companies bring together cross-functional, cross-enterprise teams to build and manage relationships with customers, partners, and suppliers in order to innovate the design, development, and delivery of products and services.

Menon and Kahn (2002) suggest that marketers should consider carefully the emotional impact of the initial encounter with a Web site, since it can affect consumers' subsequent behavior. For example, if marketers want consumers to do direct tasks on Web sites (e.g., registering, making a purchase), and deeper browsing is not required or even desired, those Web sites should be designed with interesting stimuli. Clearly, these results have implications for the design of Web sites, personalization, and online advertising.

Getting Customers to Spend Time and Interact with a Brand

A digital environment offers consumers new ways to spend time with a brand. There is no need to go to stores to see products and to talk with sales people. Over the Internet, consumers can access a vast amount of product information, view 3-D images, test different product variations, play games, and interact with other brand users. New digital environments and the emergence of multiple channels have created opportunities for marketers to be more creative than before. For example, Nokia has invented an every-year game event. In Nokia Game, players are called upon to collect and act on a series of clues delivered via a host of channels, including the Internet, the press, radio, e-mail, telephone, and television. Those that successfully put all the pieces of the puzzle together will be rewarded with a chance to participate in the grand finale. Gaming keeps consumers interacting with Nokia brand for three weeks. In 2001, there were more than 600,000 players from 28 countries (see www.nokiagame.com).

In an interactive environment such as the Web, consumers may reach a state in which their sole attention is focused solely on the universe contained within the boundaries of their network navigation experience. Consumers may comment, "I completely lost track of time." Such a state has been labeled *flow* in previous research (Hoffman & Novak, 1996). Developing games and online experiences is a new area for traditional-brand marketers, which turns around the old concept of one-way brand communication. Previous means like sponsoring and off-line events have not disappeared, but they are now more mixed with the online world. For example, some pop-artists have shown their live concerts on the Internet for free on sponsors' Web pages.

However, interactivity places demands on the structure of companies (i.e., the way they organize themselves and their processes). Firms must have proper technology, channels, and enough resources for handling customer-initiated interaction, such as feedback and inquiries. In fact, some companies have difficulties even in basic responses to customers' communications. For example, it was found that, on average, it takes 21 hours for e-tailers to respond to customers' e-mail inquiries, and 18% of the sites never respond (The DMA's State of the E-Commerce Industry Report, 2002).

Summary and Conclusion

In this chapter, we have modeled the effects of digital marketing on customer relationships. We suggest that brand communication and service are the key elements. Personalization and interactivity can be mixed into brand communication and service to reinforce the effects of digital marketing on customer relationships. Multiple channels exist for digital marketing; the Internet, e-mail, and mobile offer different options and opportunities when used separately or integrated. Digital channels allow brand communication and contacts to be more frequent at lower costs than traditional channels. Increased frequency of brand contacts has the potential to enhance brand loyalty by generating positive attitudes and emotions and brand knowledge. Personalization can be used to make more relevant and interesting brand communication. Online participation and activity by customers increases the frequency of brand contacts and boosts brand loyalty.

Online service gives customers better choices, access, control, and convenience over traditional service channels. Time and place independency with the ability of customers to actively create or shape (i.e., personalize) the form of their services builds up satisfaction and loyalty. Also, marketers can create more personal services, based on customers' behaviors and interests, which creates more positive service experiences. The interactive elements of digital media can be used for better communication in service encounters, or customers can be provided additional service when buying products, which results in more satisfied and loyal customers.

We suggest that digital marketing can have positive effects on customer relationships, which can be seen both in the short term (e.g., increased sales and response rates) and the long term (e.g., enhanced loyalty, attitudes, etc.). This conclusion can be tested empirically by using appropriate methods and metrics.

More research and empirical evidence is needed to better understand what are the most suitable forms of digital marketing that strengthen customer relationships. Ideas presented in this chapter must be further tested and improved by empirical research. While focusing on digital marketing, we recognize that marketers and customers have multiple contacts and various channels that are both digital and traditional, which are often integrated. After all, online and off-line environments are blurred in consumers' everyday lives and cannot be thought of as fully separate environments in marketing.

References

Aaker, D.A. (1996). *Building strong brands*. New York: The Free Press.

Aaker, D.A., & Joachimsthaler, E. (2000). *Brand leadership*. New York: The Free Press.

Alba, J., et al. (1997). Interactive home shopping: Consumer, retailer, and manufacturer incentives to participate in electronic marketplaces. *Journal of Marketing, 61*, 38-53.

Brondmo, H.P. (2000). *The engaged customer: The new rules of Internet direct marketing*. New York: HarperCollins Publishers.

CyberAtlas (2002). Personalization makes for satisfied consumers. Retrieved November 3, 2002, from *http://cyberatlas.internet.com/markets/retailing/article/0,,6061_989061,00.html*

Degeratu, A.M., Rangaswamy, A., & Wu, J. (2000). Consumer choice behavior in online and traditional supermarkets: The effects of brand name, price and other search attributes. *International Journal of Research in Marketing, 17*(1), 55-78.

Deighton, J. (1996). The future of interactive marketing. *Harvard Business Review, 74*, 151-160.

Gallagher, K., Foster, K.D., & Parsons, J. (2001). The medium is not the message: Advertising effectiveness and content evaluation in print and on the Web. *Journal of Advertising Research, 41*(4), 57-70.

Hoffman, D., & Novak, T. (1996). Marketing in hypermedia computer-mediated environments: Conceptual foundations. *Journal of Marketing, 60*, 50-68.

Kapferer, J.-N. (1998). *Strategic brand management*. London: Kogan Page.

Kwak, H., Fox, R.J., & Zinkhan, G.M. (2002). What products can be successfully promoted and sold via the Internet? *Journal of Advertising Research, 42*, 23-38.

McAlexander, J.H., Schouten, J.W., & Koenig, H.F. (2002). Building brand community. *Journal of Marketing, 66*, 38-54.

Menon, S., & Kahn, B. (2002). Cross-category effects of induced arousal and pleasure on the Internet shopping experience. *Journal of Retailing, 78*(1), 31-40.

Merisavo, M. (2001). *Using e-mail for relationship marketing in the consumer market: Special interest in brand attitudes and brand loyalty* [master's thesis] (in Finnish).Helsinki School of Economics.

Merisavo, M., & Raulas, M. (2004). The impact of email marketing on brand loyalty. *Journal of Product & Brand Management, 13*(6), 498-505.

Merisavo, M., Vesanen, J., & Raulas, M. (2002). *The experiences and views of Finnish professionals and managers for using digital channels in marketing* (in Finnish). Helsinki School of Economics & Swedish School of Economics.

Mohammed, R.A., Fisher, R.J., Jaworski, B.J., & Cahill, A.M. (2002). *Internet marketing: Building advantage in a network economy*. New York: McGraw-Hill.

Muniz Jr., A.M., & O'Guinn, T.C. (2001). Brand community. *Journal of Consumer Research, 27*(4), 412-432.

Pesonen, L. (2002). *Consumers' attendance in service situations in electronic channels* [master's thesis] (in Finnish). Helsinki School of Economics.

Reibstein, D.J. (2002). What attracts customers to online stores, and what keeps them coming back? *Journal of the Academy of Marketing Science, 30*(4), 465-473.

Rust, R.T., & Lemon, K.N. (2001). E-service and the consumer. *International Journal of Electronic Commerce, 5*(3), 85-101.

Schultz, D.E. (2002). Multi-channel: New term, old challenges. *Marketing News, 36*(9), 10.

Srinivasan, S.S., Anderson, R., & Ponnavolu, K. (2002). Customer loyalty in e-commerce: An exploration of its antecedents and consequences. *Journal of Retailing, 78*(1), 41-50.

Sterne, J. (2000). *Customer service on the Internet: Building relationships, increasing loyalty, and staying competitive*. New York: John Wiley & Sons Inc.

Stewart, D.W., & Pavlou, P.A. (2002). From consumer response to active consumer: Measuring the effectiveness of interactive media. *Journal of the Academy of Marketing Science, 30*(4), 376-396.

The DMA's state of the e-commerce industry report. (2002). Retrieved from *www.the-dma.org/bookstore/cgi /displaybook? product_id =009308*

Watson, R.T., Sigmund, A., & Leyland, F.P. (1998). Attractors: Building mountains in the flat landscape of the World Wide Web. *California Management Review, 40*, 36-43.

Wind, Y., Mahajan, V., & Gunther, R.E. (2002). Convergence marketing: Strategies for reaching the new hybrid consumer. NJ: Prentice Hall.

Chapter VII

Strategy Turned into Action: A Case from Global Implementation of B2B E-Business

Magnus Holmqvist
Volvo Information Technology and Viktoria Institute, Sweden

Kalevi Pessi
Göteborg University and Viktoria Institute, Sweden

Abstract

E-business has been highly debated during the last years, often based on assumptions. This chapter is based on results from several years of collaborative action research. Actual results from implementation projects in the supply chain of the automotive industry are the focus. The objective is to highlight experiences and successful results from business to business integration. The origin is a Volvo initiative with a portal for selling spare parts via a new Web channel. The case presents realized implementation projects as an outcome of a scenario-based strategy. The original scenarios served as a reference, but the development was influenced by actual actions and learnings. The case illustrates that creation of a new platform is challenging, but also that the creation of new business relations is just as difficult. The results indicate that tight follow-up projects can enable valuable innovations. Leverage is high, and rollouts are easier after the initial implementation.

Introduction: Logistics at the Core of Global Operations

The world is growing smaller. Globalization and market demands have made logistics one of the most critical business functions. Availability is the key to gaining customer loyalty. This has been the case for a long time in the automotive industry, especially in the after-market, where access to correct spare parts is decisive to the function of the vehicles. However, there also are many opportunities to utilize modern technology in order to improve the business setup (Malone, 1998). The fierce competition in the transport sector pushes the business-to-business relations to focus on bottom-line results in a reality of diminishing margins. At the same time, exploiting core competencies and finding new business propositions seem even more important. With large flows of physical goods as well as many stakeholders of information, logistics is growing into one of the most complex business functions (Ericsson, 1999, 2003; Lumsden 1998; Motwani et al., 2000). This has contributed to the growth of the Supply Chain Management area, including, for example, the emergence of both specific systems and SCM consultants.

Managing operations and changing in complex environments are problematic. Outlining strategies that involve highly dynamic factors is difficult. With multiple interrelations on one hand and large differences on the other hand, management must act in a situation where control of operations and technology is drifting (Ciborra, 2000). In a turbulent context, it may be crucial to take action. No matter how much planning, it will still not be possible to comprehend the situation in detail or to control all implementation factors (Magoulas & Pessi, 1998).

E-business often has been described as revolutionizing companies and businesses with implementations through large-scale projects (Kalakota & Robinson, 1999). However, many startups have failed due to the fact that physical operations did not deliver what their Web sites seemed to promise. There also has been a tempting suggestion that intermediates could be replaced by e-business solutions. But stakeholder relations are difficult to establish and take time to change, especially in supply-chain logistics. It is easy to underestimate the problems, risks, and challenges, which need to be managed when aligning IT and business into new channels.

This chapter presents a case of successful business-to-business (B2B) e-process implementation at Volvo, where detailed planning has been subordinated to action due to the context characteristics.

Case Background: Solutions in After-Market Logistics

Volvo and Spare Parts Logistics

AB Volvo is a world-class provider of transport solutions, services, and products. With global presence and sales exceeding 185 billion SEK, its more than 75,000 employees focus on business-to-business operations in the areas of trucks, buses, construction equipment, marine engines, and aerospace. These products are capital-intensive, and any down time is expensive, which raises the demand for spare parts availability.

The starting point of this case is a business review and development of an e-business strategy in the year 2000, initiated as an AB Volvo project. At the turn of the millennium, it was time to change Volvo's Internet solutions from being just a place on the Web, where customers could find information about products and services, to being a tool for Volvo to make direct business with customers and dealers.

A main drive for starting the initial project was a common view of the whole automotive industry that the Internet, as a new technology, enabled a huge opportunity to create a new efficient channel to reach and to make business with end-customers. End-customers are vehicle owners and/or operators. However, large variations exist in terms of size of business, geographical location, and type of operation (e.g., pan-European freight fleets, single drivers, and city bus operators). For Volvo, three main incentives were identified, which justified actions in order to get closer to end-customers:

- **Competitors:** Other truck manufacturers or new entrants may build new channels. Special focus on possible third-party Internet companies that wanted to sell spare parts. This potentially was a large threat for the after-market business of Volvo.
- **Cost Reductions:** The Internet and e-business could increase productivity in existing processes and improve support given to end-customers but especially to dealers with services like online training for mechanics, information and document distribution, and spare part sales.

Figure 1. Volvo parts information and goods flow (Volvo Parts Operational Guide, 1998)

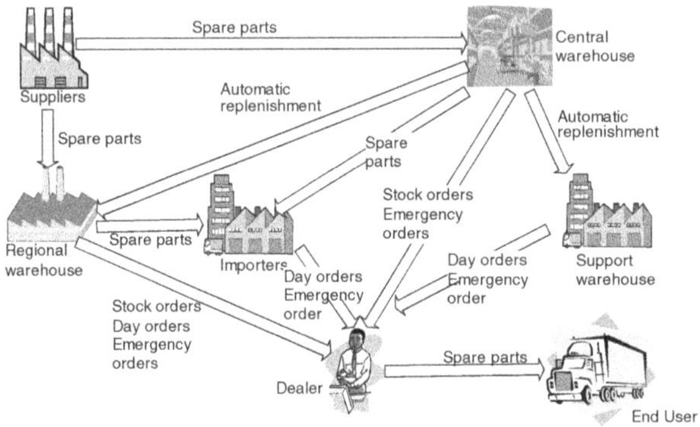

- **New Services:** To broaden and expand the total offer; simplified telematic services, load matching, spare parts purchase.

This case will focus on after-market business with spare parts logistics as the core operation. Spare parts logistics is a complex operation characterized by intensive physical and information exchange between several stakeholders; in the case of Volvo, thousands of suppliers and ten thousands of distribution points toward hundreds of thousands of end-customers. Every day (365 & 24/7), around 70,000 order lines are handled for more than 185 markets. The industrial product families contain hundreds of thousands of parts, which demand handling of both a long-term service responsibility and complicated super-session chains. The parts also increase in complexity, as they are no longer just physical but also digital and part of service arrangements and wider business solutions, as well. In Figure 1, the distribution of the parts is depicted; it is not necessary (and in many cases, it does not happen) that the information flow goes the same way (the end-customer is referred to as end-user in the figure).

Thirty Years of Evolutional Business Development and Systems Integration

In order to understand the context of the case in this chapter, it is essential to have a brief background of the supply-chain evolution in Volvo. During the 1970s, when the usage of information systems started to evolve, Volvo already had more than 40 years of experience in manufacturing vehicles. The after-market population, both in terms of population size and range, was considerable. Hence, as computerization possibilities emerged, one of the first areas in which to reap benefits was control of stock balances in order to maintain quality. A clear breakpoint for the development was the introduction of mainframe technology, although in a limited area. Automation was the real objective. Instead of keeping logbooks, doing manual recounts, awaiting status, and so forth, the first computer programs could release resources at headquarters, giving possibilities to further expand and internationalize. At the beginning, the internationalization was mainly ad hoc setups of an entrepreneurial spirit, which helped establishing market presence. However, these have gradually transformed into organizations with sophisticated systems for after-market services. Largely due to utilizing IT, it became possible (from a headquarters' perspective) to embrace and integrate information further out in the supply chain (i.e., closer to the original information source).

The fast buildup of international establishments in the 1980s had created diversity and decentralization. Systems to handle the after-market business were built up to an importer level. These were decentralized systems based on client/server technology. Still, relationships grew stronger between spare parts importers and headquarters, which was a prerequisite for gradual system and supply-chain integration.

The fast growth, together with a tougher business climate, had made times right during the early 1990s for larger rationalizations and improved control at Volvo. Trends of both BPR and ERP (Enterprise Resource Planning) systems were dominating the scene. But neither BPR nor ERP contributed to radical changes at Volvo, despite the fact that these trends often were presented to revolutionize the business. However, ERP modules from SAP R/3 were introduced in one of the non-core processes. It was beneficial for the accounting function. At the same time, international relations were maturing toward the importers. Common ways of working could be introduced with one system consolidating the whole chain; however, just with regard to financial transactions. At the same time, management drove both overall business consolidation and system integration with suppliers and emerged toward dealers. With suppliers, it was mainly through EDI, and with dealers, it was through practical cooperation. Concepts of the extended enterprise emerged (e.g., through introducing the practice of Vendor Managed Inventory (VMI).

Globalization and mergers have been characteristic of the late 1990s, combined with an upswing for innovations and utilization of communication technologies. In the case of Volvo, cooperation with Mitsubishi, Schmitz, and Renault/Mack has either been intensified into mergers or disinvested.

Over the past 30 years, one can discern an evolution toward a large and global company with growth phases leading to diversification but in certain areas again becoming centralized. It has been an evolution especially by means of an increasingly integrated supply chain, as conceptually illustrated in Figure 2. It depicts how Volvo has evolved, focusing on the outbound relations. Of course, inbound relations toward suppliers also have developed with more than 80% of purchase value having been handled electronically already in the 1980s and currently reaching above 95%.

The first system platform (mainframe) originally just treated internal headquarters transactions. However, importers were gradually integrated into an in-house developed logistics system. This consisted of decentralized client/server systems connected to the central platform. With better process understanding and relations as well as joint systems with many partners, an Extended Enterprise arose. In some areas, there also was integration to certain ERP modules. The current spread and adoption of Web technologies have provided opportunities for further networking and integration along the supply chain toward the end customer. However, expectations from end customers are high, caused by the broad reach of the universal usage of the Internet. With improved access and information availability, new requirements of support and response times occur. For example, end customers more commonly expect round-the-clock service. As opportunities increase in terms of availability, there are also new demands. De facto standards and transparency between platforms are mainly positive, but there are also practical obstacles with distributed technologies (e.g., scalability, load balancing, and interoperability). Getting the customer order point closer to the source of the demand gives rationalization gains through getting information input at the source as well as automatic inquiry possibilities.

Earlier system rollouts were time consuming due to the fact that they made point-to-point to each location. By introducing Web technologies, such obstacles can be eliminated, but end customers may experience problems due to differences

Figure 2. Integration toward end-customers

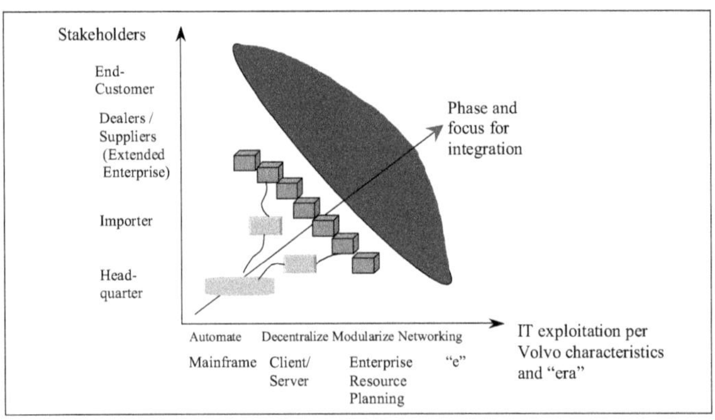

in workflow and/or infrastructure (e.g., message broker, browser engines, character recognition, connection speed/reliability, and security protocols).

Characteristics that have been highlighted in this case background were all part of the general awareness and input to the business review and e-business strategy.

Scenario Development

The desire and possibilities to integrate toward end customers has been a process characterized by evolution. At the beginning of this millennium, Internet and Web technologies were already established, but direct business relations between Volvo and end customers were not obvious. However, a strategic review of opportunities and threats led to different alternatives.

In order to find what the first step would be and how to outline a frame of reference, different scenarios were developed. Scenario development was used as an alternative to traditional strategic planning (Ratcliffe, 2002). There was an awareness that since many things were not possible to control, it would be essential to learn from implementations. But at the same time, the scenario development would provide both a holistic view and a set of business assumptions to refer to. Four scenarios where developed for the imminent introduction of online services Web portals, including spare parts functionality (see Figure 3). The Today scenario reflected the fact that there were no established e-business channels toward end customers. The developed scenarios were described in detail as well as how the relations between the stakeholders in the supply chain could be affected (Holmqvist et al., 2001, 2003).

Scenario 1. The spare parts are distributed directly from the support ware house to the end customer, which is the customer's workshop in all scenarios.

Figure 3. Four scenarios for spare parts distribution (Holmqvist et. al., 2001)

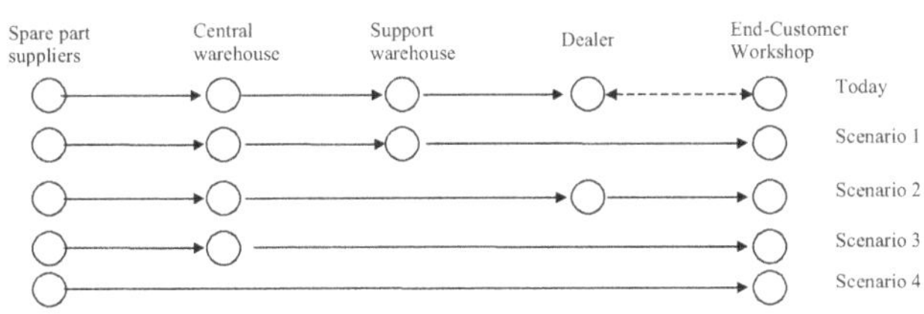

Scenario 2. The spare parts are distributed directly from the central ware house to the dealer and then from the dealer to the end customer.

Scenario 3. The spare parts are distributed directly from the central ware house to the end customer, bypassing both support warehouse and dealer.

Scenario 4. The spare parts are distributed directly from the supplier to the end customer, bypassing all traditional distribution centers.

The four scenarios analyzed the physical flow in order to determine lead times, costs, and practicalities connected to the distribution of spare parts. Furthermore, policies and business logic were evaluated in order to design e-business solutions that allowed the information flow as well as the financial flow to take other routes. For example, it would be possible for all involved actors to share certain information and enable different solutions for different spare part numbers. The scenarios also display a generalized picture of the complex structure, which actually includes minor flows and geographical alternatives (Figure 1 depicts the distribution model as described by Volvo). Fundamentally, the closer to the central warehouse you get, the larger the volumes will be and the more automated the information flow can be. Similarly, further out in the supply chain, there is a more heterogeneous structure with smaller volumes and volatile demands. Nevertheless, through the whole chain, the focus remains on availability. Relations are well founded, and the introduction of new solutions has to be introduced with precaution.

Volvo has been working on implementing concept offers and Web services (in accordance with the creation of value chain scenarios) in order to meet different requirements, thus being able to differentiate the level of integration, depending upon relationship.

Parts Online: What Became Real?

The developed scenarios and an initial implementation included a platform for Web portals that was aimed at dealers, but it also embraced a new target group—end customers. The portals include services on the Internet that support the customers' and dealers' total business cycle (e.g., getting information about services and products, operating and following up of vehicles, ordering spare parts, and reselling old vehicles). A public part of the portal can be visited at www.volvotrucks.com/onlineservices. This case focuses on service for spare parts, called Parts Online.

For Parts Online, the original goal was to increase sales to end customers with accessories and consumable parts as well as spare parts to customers operating their own workshop. Parts Online is a user-friendly system, where customers can search for and order spare parts 24 hours a day, seven days a week. Gains also are made through productivity increases and cost reductions in the form of less work for service support and help desks (traditionally, a lot of time is spent on the phone with customers). As assumed, it has been a large issue to manage relationships between dealers, customers, and Volvo. Focus has been on creating a win-win-win situation rather than on eliminating any actor in the traditional value chain. Instead, the overall aim has been to increase competitiveness of the total setup and improve supply chain productivity.

A main obstacle was to reassure the dealers that the solution was built to support their business and not to bypass them. An example to overcome this was initially to let customers register, make the business agreement with a dealer, and buy spare parts directly from them. The solution (see Figure 4) originally included the following main functionality:

- Parts catalogs
- Your favorite parts
- Direct order entry

Figure 4. Example of a Web portal screen in Parts Online

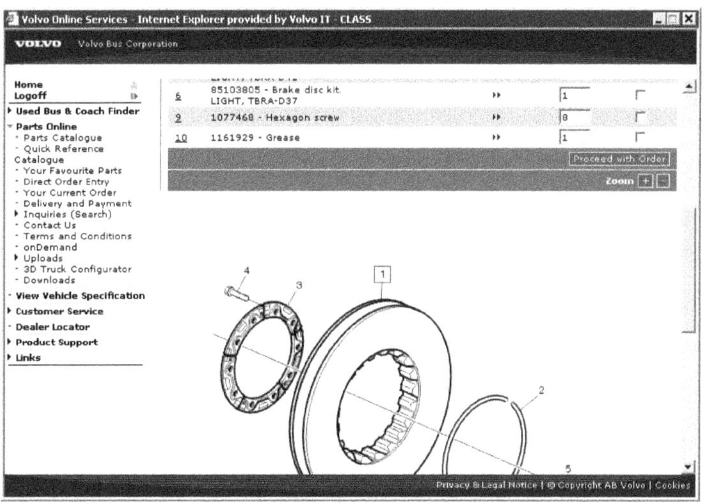

- Order status
- Inquiries (search) (e.g. parts info)

The first implementation project to be deployed strengthened the relationship between the dealer and the end customer by providing an additional channel for spare parts. Volvo also could strengthen the relation with its dealers by improving the dealer offer and supply-chain performance. The ultimate goal and strategy were to attract end customers to the dealer workshops. Referring to these scenarios, it can be described as building upon Scenario 2, where the main impact was commercial relations, the technical platform, and extending the supply chain with delivery options, as well as ordering systems to end customers.

The first implementation has enhanced Volvo's awareness of the dealer situation and enabled dealers to have online contact with end customers. It challenged existing relations through the setup of a new channel, but it was not revolutionary in any way (see Figure 5). Nevertheless, the first implementation was valuable, at least in order to keep momentum through actions. This contributed to practical insight of different business opportunities leading to follow-up projects.

The learning facilitated the next implementation project. The platform was easily extended in version 2 to deliver enhanced technological features to established relations (e.g., independent importers). In some special cases, it also established business directly with end customers. Still, involving the close proximity and human relation of the dealers, the system and logistics connection goes from a

Figure 5. Version 1. Implementing Scenario 2—reaching end customers

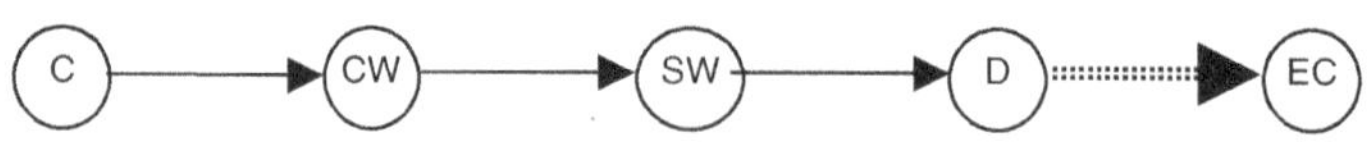

Figure 6. Version 2. Restructuring relations – Scenario 3

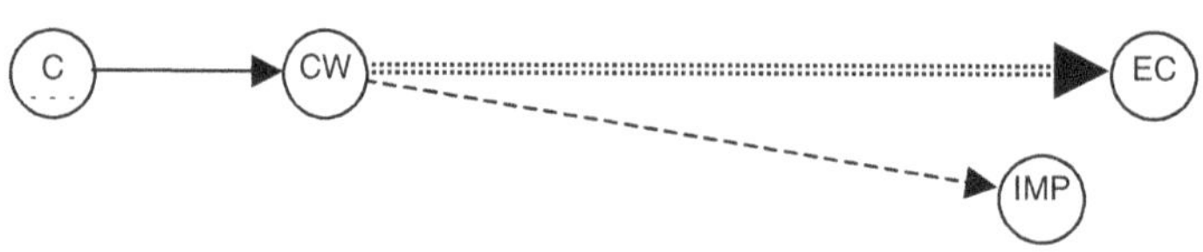

Figure 7. Version 3. Increase globalisation – Scenario 1

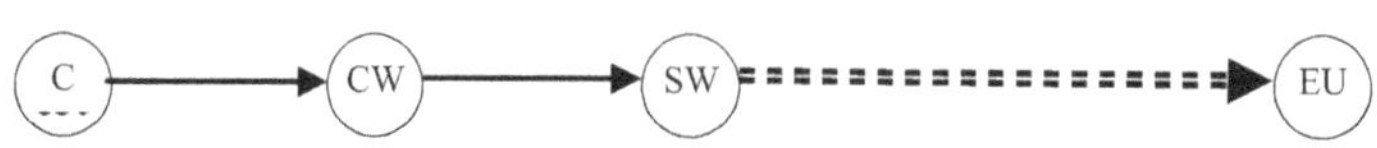

central warehouse to end customers. This setup only would be valid if certain criteria were fulfilled, such as that end customers have and will maintain their own workshop over a foreseeable time and as long as distribution points and volumes are justifiable. This relates to the initial Scenario 3 and also to an innovative win-win-win situation for end customers, dealers, and Volvo.

Further implementation projects extended reach and increased globalisation. This was somewhat a realization of Scenario 1 from the scenario development by extending the capabilities of support warehouses to deliver to end customers. Technically, the platform needed minor adaptations, and the objective is rather to extend logistical capabilities (e.g., in East Europe and Asia). At the same time, obstacles of the more decentralized setup had to be overcome (i.e., several support warehouses) and, therefore, required close analysis and monitoring during rollout. Previous experiences proved very valuable for the success of the implementation.

The overall impression of the development is that work that was made during version 1 gave payback in the following versions. Both the cost as well as the lead time were lowered in later versions. From version 1, a lot of experience regarding

Table 1. Summarized view of implementation projects

	Original Implementation	**Follow-up Project 1** (Version 2)	**Follow-up Project 2** (Version 3)
Development Cost (relative size)	1	1/5	<1/20
Lead time (relative)	1	½	1/4
Challenges			
-Project Process	Complex, many functions involved – but good focus and management attention	Easier, since follow-up project with benefits in learning curve	OK, but more difficult, since decentralized and less focus due to less management attention
-Technology	New, Complex	Stabilizing (minor challenges)	Maturing (but diffusion emerging)
-Roll-out	New setup, piloting	Fast upgrade, good focus	Moderate with many actors
-Commercial Relations	Existing but new channel and collaboration agenda	Emerging channel and new relations	Experience from former setups
-Physical Logistics	Not focused	New setup and flow	Utilizing the established setup
Results			
-Cost/Benefit	Emerging benefits	Great benefits	Benefits as expected
-Innovation	New channel	Win-win-win relations	Extended reach and features, re-use of Web services

Internet technology, commercial relations, and logistics setup was gained. The tight follow-up projects gave innovations in terms of usage and setup for physical distribution, financial flows, and design of information availability. In particular, enhanced information availability contributed to a widespread opinion of being easy to do business with. Furthermore, a parallel and increasingly important factor is Web services deployments. Actual implementation experiences and results from advanced Web services are scarce so far. The commercial value and interest are vivid. For example, IBM is using one of the implementations as a case study for Web services and is presenting it in its company-of-the-month section (see http://www-306.ibm.com/software/ebusiness/jstart/casestudies/volvo.shtml. It shares some of the project work, although it is a simplified and commercialized view of the objective, implementation, and even the origin of the overall developments. The continuous actions can be viewed as a summary of the development and implementation results in Table 1.

Concluding Remarks

This chapter presented a case of global implementation of B2B in after-market e-processes. In global spare parts logistics, we found that large volumes and a considerable installed base (dependent upon economies of scale) have the pressure to improve performance as well as to be innovative. Still, with such facts, albeit industry maturity and competition, it is difficult to implement radical changes. Instead, it is important to apply opportunities where they fit and are applicable rather than to believe that there are everything/everywhere solutions. Development projects need a size that is possible to comprehend and, in that way, increase steering, as scope may drift due to different stakeholder objectives. Having several implementation projects with clear target groups that can reuse platform and Web services is clearly beneficial. Evolutional development seems to require persistence and an ability to handle mismatch between IS/IT and business structures. Even considering that it takes time to change, it is equally important to regard and assess how different actors may act upon available opportunities.

The most obvious benefit of Parts Online is providing users with information availability; that is, always updated and online data, increased productivity, time saved, single log on, 24/7, and being easy to do business-with. However, this will be worthless if the physical logistics fail to deliver the desired performance or if relations between the stakeholders are perceived as a threat to business. Volvo has gained benefits in terms of increased customer satisfaction and loyalty, increased productivity (i.e., less questions, errors, double data input), and lower development costs (by shared development between different versions and set-up).

The new channel for B2B e-processes at Volvo relies on the development of integration between legacy and a new IT platform, as well as the development of relationships among stakeholders. The case has shown that continuous actions and step-by-step implementation projects on a solid platform bring innovations that have enabled a new business structure. Specifically, continuous implementation projects can deliver innovation in new relations and new channels. The leverage of these innovations is high, and it is easier to roll out the new channels after the first implementation.

References

Ciborra, C. (Ed.). (2000). *From control to drift. Monteiro E. Actor-network Theory*. Oxford: Oxford University Press.

Ericsson, D. (1996). *Virtual integration*. Norcross, Canada: Unisource.

Ericsson, D. (2003). Supply/demand chain management—The next frontier for competitiveness. In D. Waters (Ed.), *Global logistics*. London: Kogan Page.

Holmqvist, M., Hultkrantz, O., Stefansson, G., & Wingqvist, A. (2001). Consequences of e-commerce on physical logistics—A theoretical scenario for spare part distribution. *Proceedings of the 9th WCTR Conference*, Seoul, Korea.

Holmqvist, M., Hultkrantz, O., Stefansso, G., & Wingqvist, A. (2003). Consequences of e-commerce for spare part distribution. *Proceedings of LRN Conference*, London, UK.

Kalakota, R., & Robinson, M. (1999). *E-business: A roadmap for success*. Addison Wesley.

Lumsden, K. (1998). *Logistikens Grunder* (Foundations of logistics). Lund, Sweden: Studentlitteratur.

Magoulas, T., & Pessi, T. (1998). *Strategisk IT-management*. Doctoral thesis. Göteborg, Sweden, Göteborg University.

Malone, T., & Laubacher, M. (1998). Dawn of e-lance economy. *Harvard Business Review*.

Motwani, J., Madan, M., & Gunasekaran, A. (2000). An information technology in managing global supply chains. *Logistics Information Management, 13*(5), 320-327).

Ratcliffe, J. (2002). *Scenario planning: Strategic interviews and conversations*. Bradford, UK: Emerald Group Publishing Limited.

Chapter VIII

Opportunities and Challenges for B2B Manufacturing Firms: Moving from Products to Services-Case SKF

Esko Penttinen, Helsinki School of Economics, Finland

Timo Saarinen, Helsinki School of Economics, Finland

Abstract

Today, many traditional manufacturing firms are focusing on their service operations, which are often seen as a better source of revenue than the first-time installations. Information and communications technology (ICT) can accelerate this process by offering efficient ways to deliver services to the customers and by allowing companies to transform their traditional product strategies to services. This chapter tells the story of a traditional component supplier that wanted to become a service firm. The transition is analyzed based on an established theoretical framework identifying efficient strategies for delivering different types of services, thus providing managers with guidelines for choosing the appropriate business model. The findings of this single case study confirm that the role of ICT as an enabler in the transition process is significant. Furthermore, the buyer's trust in the supplier turned out to be an essential factor in developing new service-based business.

Introduction

Levitt (1972) first announced that everybody is in service. It has taken a long time for companies and researchers to make sense of this provocation. It was not until recently that researchers turned to manufacturing companies and found that manufacturing strategy can be successfully based on a service approach (Fry et al., 1994). Today, managing the transition from products to services is a very important emerging managerial issue in the business-to-business environment (Oliva & Kallenberg, 2003).

The objective of this chapter is to describe how one specific firm—Svenska Kullager Fabriken, SKF—has undergone the change from product-oriented thinking to services-based concepts. We analyze this transition using an established framework and discuss the challenges and opportunities that the repositioning of the company's offering creates. More specifically, we use the theory of full-service contracts (Stremersch et al., 2001) to describe how SKF has moved from offering products to offering full-service contracts. In addition, we use the Service Process Analysis (SPA) (Tinnila & Vepsalainen, 1995) to evaluate the different channels that SKF uses for the different kinds of offerings.

We find that the information and communications technology (ICT) provides firms with a strategic tool that acts as the main enabler in the transition from products to services. In the case of SKF, information technology is used via endorsia.com to build closer electronic linkages to customer companies. In addition, SKF has added intelligence to its products by developing enhanced status-reporting bearings. These two ICT developments—endorsia.com and intelligent bearings—are described in this chapter. Also, we find that the buyer's trust in the supplier is an essential factor in the transition from products to services-based systems. The finding is in line with previous theoretical work, which suggests that these enhanced relations between a buyer and a seller require considerable trust in order to create the full-service offering.

The chapter is organized as follows: the next section briefly presents the theories of full-service contracts and service process analysis. The third section introduces the case SKF. In the remaining sections, we present theoretical and managerial implications.

Full-Service Contracts

The traditional marketing literature defines services as intangibles, variables, and perishables. Therefore, the consumption and production processes of services cannot be separated. Rust et al. (1996) and Kotler (1999) define products as "anything that can be offered to a market for attention, acquisition, use, or consumption that might satisfy a want or need. It includes physical objects, services, persons, places, organizations, and ideas" (p. 291). A service is seen by Kotler (1999) as "any activity or benefit that one party can offer to another that is essentially intangible and does not result in the ownership of anything" (p. 291).

In our study, we use the concept of full-service contracts (Stremersch et al., 2001) to examine the transition from product-oriented thinking to services-based concepts. Full service is defined as a bundled offering that is an extension in meeting customer needs, requiring interaction between the supplier and the customer and thus making it more challenging than traditional delivery (Figure 1).

The bundling literature, initiated by Burstein (1960) and Stigler (1963) and later formally formulated by Adams and Yellen (1976) originally seeks to contemplate why firms often sell their goods in packages (e.g., sporting and cultural organizations offer season tickets and restaurants provide complete dinners). In the business-to-business context, bundling is often used to create full-service offerings in order to provide customer companies a single point of contact (Cristol &Sealey, 1996; Stremersch et al., 2001). A review of current literature on bundling is provided in Stremersch and Tellis (2002) and Penttinen (2004).

The second dimension, extension in customer needs, describes the extent to which customer needs are satisfied by the supplier firm. The three levels are single, extended, and total need fulfillment. The general proposition of many academic and practitioner-oriented research papers has been that firms are developing more comprehensive solutions and services to meet customer needs (Goldman et al., 1995; Weill & Vitale, 2001; Wise & Baumgartner, 1999).

Figure 1. Full-service contracts (Stremersch et al., 2001)

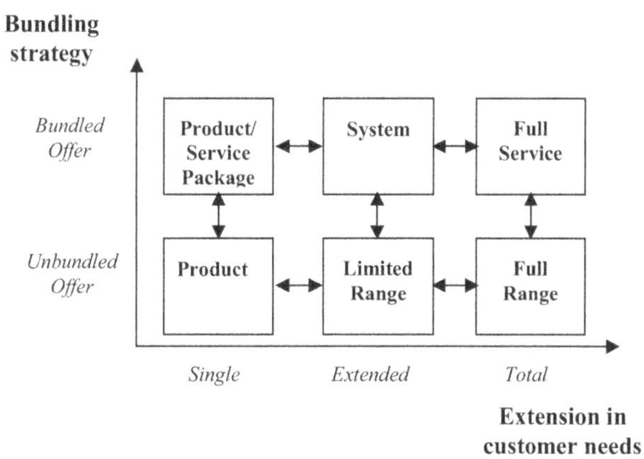

In order to investigate the different delivery channels within this transformation from products to services, we need a theoretical framework that allows us to match different types of channels to different types of services. For this purpose, we use the Service Process Analysis (SPA) (Tinnila & Vepsalainen, 1995). The objective of the SPA is to evaluate the efficiency of matching the type of service being offered with alternative types of delivery channels. The axes of the SPA matrix, therefore, represent the type of service and the type of channel. The three types of channels used in this study are agent, service personnel, and market network. The three types of services are customized delivery, standard contract, and mass transaction. The efficient combinations can be found on the diagonal. (1) Focused processes should be used for customized services offered by specialized agents. (2) Flexible integrated processes can be used to handle standard contracts offered by service personnel. (3) Mass transaction services can be handled via market network through fast routine processes. Figure 2 illustrates the framework.

The SPA model facilitates the analysis of strategic repositioning of services, such as replacing the service provided by a new type of service, resegmentation of a full service into several specialized services, or establishing a new channel. Advances in data networks have created several new types of channels providing services discussed mainly in the information systems literature (Tinnila, 1997).

Figure 2. Service process analysis (adapted from Tinnila & Vepsalainen, 1995)

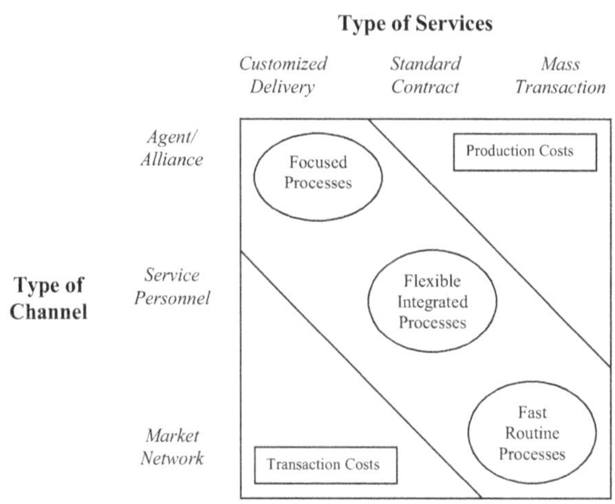

Case SKF

SKF (Svenska Kullager Fabriken) is the leading global supplier of bearings. SKF was founded in 1907 and, from the very beginning, focused intensively on quality, technical development, and marketing. Today, the company employs 38,600 people worldwide and has a turnover of about 4.5 billion euros. The company produces bearings, which are used in all kinds of rotating machines and equipment. This chapter presents the case of SKF Finland. The biggest bearing users for SKF in Finland are the paper and pulp industry as well as the steel industry, but, in addition, all the machine manufacturers of gearboxes, electrical motors, pumps, and so forth are large consumers of bearings. Globally, the biggest bearing user is the vehicle industry.

Bearings are used to keep rotating shafts in their position, to ease friction between moving parts, and to carry load. Bearings are produced in various sizes and shapes, with the smallest bearing weighing just a few grams and the larger ones weighing a few tons. The material used to produce bearings can vary from conventional steel to ceramics, brass, and plastics. Roller bearings are mainly made of special steel.

Recently, with the help of the strong application and maintenance knowledge of bearing mounting, lubrication, condition monitoring, and so forth, the company has begun offering comprehensive maintenance contracts to its customers. The electronic marketplace and the intelligent bearings already described play a key role in this new strategy of SKF.

Moving Toward Full-Service Contracts

Throughout the 1980s, SKF has been seen on the market as a high-quality component manufacturer that provides customers products and services on a transactional basis. Products and related services were sold and purchased separately; they were not bundled together. The customers interacted with SKF on issues related to logistics, product development, and the like, but the operational linkages and information exchange were relatively low.

As a response to the increased customer demand, SKF extended its product line by developing new products and related services and by bundling them together. The innovative bearings increased the production output capacity of one customer's paper machine, and the bundled purchasing of related material (i.e., lubrication oil) made life easier for another.

In 2000, the company began offering a totally new kind of business deal—runability and operability of bearings. In the new business offering, SKF guaranteed to keep the bearings rolling in the customer's machines.

> *The aim of SKF Integrated Maintenance Solution is to offer customers a service which would make it possible for the customer to outsource their bearing maintenance, i.e. SKF would take the full responsibility of the bearing maintenance including condition monitoring, lubrication, replacement and logistics needed to get a new bearing for replacement. The customers of SKF are invited to outsource everything related to bearings to SKF. This would mean that the customers of SKF would no longer have to purchase the bearings, they would in fact purchase simply comprehensive maintenance functions and runability for their machinery.* (CEO SKF Finland)

Each agreement is customized to specific business needs. The customer can choose which areas it wants to include, based on internal resources and current supplier contracts. "These new concepts require SKF to build closer relationships with the customers and to understand well the processes of the customers" (CEO SKF Finland). Within a full-service contract, SKF plays a significant part in increasing the reliability and integrity of the plant. SKF shares some of the risk as well as the savings, while the customers receive agreed-upon financial returns without capital investment.

Therefore, in response to the changes taking place in the industry, SKF has developed new products and service concepts to offer to its customers. These new concepts require SKF to build closer relationships with the customers and to better understand the processes of the customers. Two different concepts that illustrate this are endorsia.com and intelligent bearings, which are part of the new SKF service strategy. The following figure illustrates the move from single unbundled product offering toward full-service bundled offering.

The company first extended its product range to include additional products and services that support the core product (bearing). Then, in the second phase, SKF bundled these products and services together and started to offer operability of bearings as a full-service concept (Figure 3).

Setting the Stage: endorsia.com and Intelligent Bearings

Next, we will discuss two information technology components that SKF has implemented in order to offer the full-service contracts: (1) the creation of endorsia.com, an electronic marketplace that facilitates the connections between SKF and its clients by providing a single interface and a platform for doing business and (2) the development of intelligent bearings, which are an important part of the new SKF service strategy.

In order to streamline the purchasing processes toward their customers, SKF, INA, Timken, Sandvik, and Rockwell Automation came together to create an electronic marketplace called *endorsia.com*. These five companies are the owners of the marketplace, but endorsia.com is open to all brands that sell industrial goods. One of the main advantages of endorsia.com is that it brings together large numbers of buyers and sellers. Therefore, the products bought and sold in the marketplace are not limited to bearings. The marketplace gives users the opportunity to connect with their preferred suppliers and customers at a single interface and to create a platform for doing business. For distributors and end users, endorsia.com promises an efficient and cost-effective purchasing system. The user can use this single interface to access real-time technical information, product availability, delivery times, and conditions.

Figure 3. Moving toward full-service, case SKF

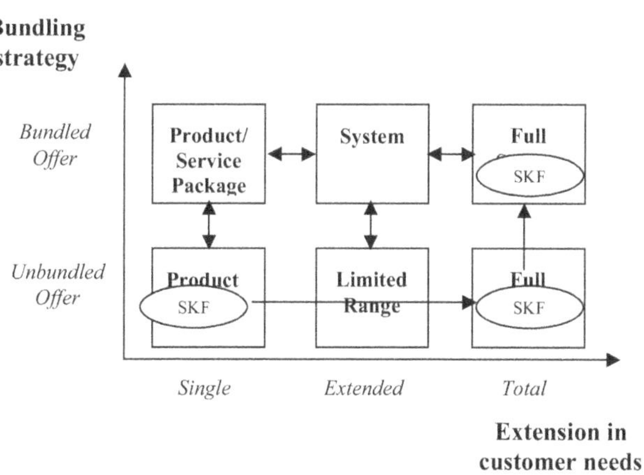

In addition to creating this marketplace, SKF has developed intelligent bearings. With this new type of bearings, the user can get more accurate data on the condition of the bearing by measuring the run-out and the vibration of the bearing. This can be done with the help of a sensor that is attached to the bearing or the bearing housing. Naturally, there are considerable benefits of knowing more accurately the status of the bearing condition; for example, there is an hourly monetary loss of more than 10,000 euros when a paper machine cannot be used due to a mechanical problem (i.e., a bearing failure). SKF offers its clients software to be used with these intelligent bearings. The software makes a connection between the client's ERP (enterprise resource planning) systems and endorsia.com. For example, this allows the replenishment processes to be handled automatically.

Discussion

Theoretical Implications: Differentiated Offerings

This chapter discusses the move toward full-service contracts. However, SKF also offers component sales for those customers not willing to invest in the new business offering. The objective of the component sales is cost minimization, whereas the full-service contracts aim at improving plant productivity. To compare these two different offerings (i.e., component sales and full-service contracts) to the SKF's initial positioning, we use the service process analysis matrix.

We can see that SKF has differentiated its positioning on the SPA matrix by creating two different offerings. There are component sales (B) for those clients who are interested in cost minimization and full-service contracts (A) for those who are willing to work in close collaboration with SKF (Figure 4). This finding of general services differentiating toward expert services and, in addition, toward mass transaction services is supported in earlier research (Kemppainen & Vepsalainen, 2003; Tinnila & Vepsalainen, 1995).

Managerial Implications: IT and Trust

The industrial maintenance business is undergoing major changes. Many maintenance supplier companies see their role extending toward the client company's processes. This chapter has described how one specific company—SKF—has

Figure 4. SKF's differentiated offerings

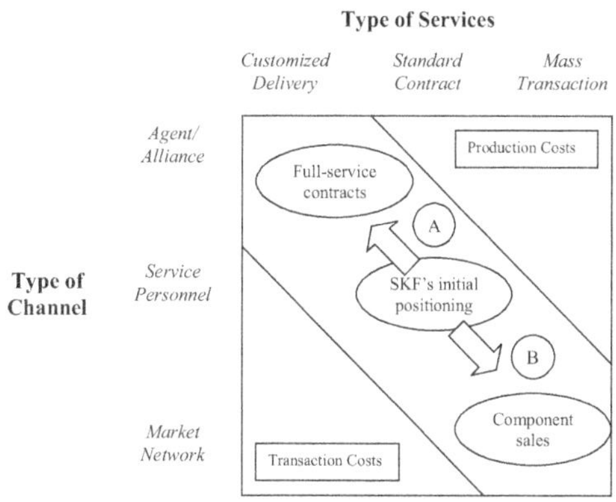

differentiated its product/service offering. Figure 5 presents the evolution of maintenance from component sales toward partnerships.

From case SKF, we can see that the trust in the supplier is a key factor in the move toward full-service contracts. Full-service contracts require some responsibility to be shifted from the customer to the supplier, and this naturally requires significant trust in the supplier on the client's part.

Figure 5. Evolution of maintenance at SKF

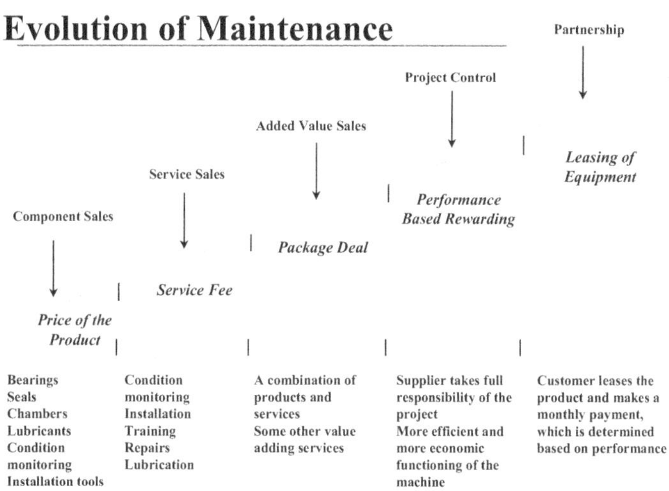

In addition to the trust, investments in innovative information technology played an important part in the transition toward full-service contracts at SKF. It would be impossible for SKF to offer these contracts without the intelligent bearings and the electronic marketplace, endorsia.com.

Conclusion

This chapter tells the story of a product manufacturer that first began offering additional services to support the product and then proceeded to offering full-service contracts to those customers who were interested in close collaboration and partnerships. This single-case study reveals the role of information technology as the main enabler in the transition. Also, the customer's trust in the supplier plays an important role in the change from product sales toward full service. Therefore, a company contemplating the move toward services-based concepts should invest in innovative information and communications technology and remember that building trust is very important.

References

Adams, W., & Yellen, J. (1976). Commodity bundling and the burden of monopoly. *Quarterly Journal of Economics, 90*, 475.

Burstein, M.L. (1960). The economics of tie-in sales. *Review of Economics & Statistics, 42*, 68-73.

Cristol, S.M., & Sealey, P. (1996). Replacement marketing. *Marketing Management, 5*,12-23.

Fry, T.D., Steele, D.C., & Saladin, B.A. (1994). A service-oriented manufacturing strategy. *International Journal of Operations & Production Management, 14*, 17.

Goldman, S., Nagel, R., & Preiss. K. (1995). *Agile competitors and virtual organizations—Strategies for enriching the customer*. New York: Van Nostrand Reinhold.

Kemppainen, K., & Vepsalainen, A.P.J. (2003). Web breeds services apart—But how to get them right? In T. Reponen (Ed.), *Information technology-enabled global customer service.* Hershey, PA: Idea Group Publishing.

Kotler, P. (1999). *Principles of marketing*. London: Prentice Hall Europe.

Levitt, T. (1972). Production-line approach to service. *Harvard Business Review, 50*, 41.

Oliva, R., & Kallenberg, R. (2003). Managing the transition from products to services. *International Journal of Service Industry Management, 14*, 160.

Penttinen, E. (2004). *Bundling of information goods—Past, present and future* [working paper]. Helsinki, Finland: Helsinki School of Economics.

Rust, R., Zahorik, A., & Keiningham, T. (1996). *Service marketing*. New York: Harper Collins College Publishers.

Stigler, G.J. (1963). United States v. Loew's Ink: A note on block booking. *Supreme Court Review, 152.*

Stremersch, S., & Tellis, G.J. (2002). Strategic bundling of products and prices: A new synthesis for marketing. *Journal of Marketing, 66*, 55.

Stremersch, S., Wuyts, S., & Frambach, R. (2001). The purchasing of full-service contracts: An exploratory study within the industrial maintenance market. *Industrial Marketing Management, 30*, 1-12.

Tinnila, M. (1997). *Division of service and business processes* [doctoral dissertation]. Helsinki, Finland: Helsinki School of Economics.

Tinnila, M., & Vepsalainen A.P.J. (1995). A model for strategic repositioning of service processes. *International Journal of Service Industry Management, 6*, 57.

Weill, P., & Vitale, M. (2001). *Place to space—Migrating to eBusiness models*. Boston: Harvard Business School Press.

Wise, R., & Baumgartner, P. (1999). Go downstream: The new profit imperative in manufacturing. *Harvard Business Review, 77*, 133.

Chapter IX

Shifting Perspective from Design to Business:
Extending the Scope from Spectators to Stakeholders

Andreas Nilsson, Viktoria Institute, Sweden

Abstract

This chapter investigates potential business value based on the results of a design-oriented research project. The objective of the research was to design mobile applications and services that support spectators in situ *at a trotting track. Ethnographic field studies have guided the development of the system TrottingPal. The research has conducted system evaluation through two field trials in the authentic setting with target users. By extending the scope to relate also to other stakeholders in this setting, additional benefits from the TrottingPal system are revealed. This chapter suggests that the designed technology targeted at spectators does not offer only a potential business relation between an event organizer and spectators; rather, it offers new channels for customer relations such as sponsorship and strengthens the locality at the track.*

Introduction

During recent years, scholars have drawn the contours and anticipated the next wave of mobile computing (e.g., Lyytinen & Yoo, 2002). This has been described as being facilitated by increased technology support while being on the move and having improved bandwidth and reliability of wireless networks. It has been claimed that new business models and value chains can be developed accordingly to accommodate novel mobile support and services (Varshney & Vetter, 2001). Despite high expectations of third generation cellular networks (3G), several European investors and telecommunication companies have noticeably bided their time to await market reaction on pioneering projects and ventures. During 1998, m-commerce revenues were forecast to be $2 billion in 2002, but stopped at $500 million (Urbaczewski, Valacich & Jessup, 2003). Accordingly, it has been argued that the missing piece is an understanding of the motivations and circumstances surrounding mobile device use and adoption from the perspective of the consumers themselves (Sarker & Wells, 2003).

This chapter investigates potential business value of the TrottingPal system. However, to be able to discern possible business potential, one cannot only view the resulting application as the starting point for commercial value. Rather, the scope needs to relate to stakeholder goals. By investigating stakeholder goals, the benefits within the value chain as a whole can be examined while TrottingPal is used. The hypothesis is that potential business possibility exists with TrottingPal. The contribution of this chapter is twofold. First, it provides a critical investigation of migrating research results to a business possibility. Second, it provides a strategic discussion of the benefits of TrottingPal for the stakeholders. The previous research on TrottingPal (Nilsson, 2004; Nilsson & Nuldén, 2003) has focused solely on the spectator without looking at potential benefits for other stakeholders.

The chapter begins with a background that outlines the betting business in Sweden, the previous research on developing TrottingPal, and related work. The background on the betting business raises a set of emerging problems. The chapter then turns to an analysis, which revisits the research, adding a stakeholder perspective. By revisiting the previous work on TrottingPal and while adding a stakeholder perspective, the chapter looks at ways in which the implementation of mobile technology (i.e., TrottingPal) can meet and accommodate the emerging problems. This is followed by a discussion before the chapter concludes.

Background

This section describes the betting business in Sweden with an outline of emerging problems, previous research on developing TrottingPal, and a brief literature review.

The Betting Business in Sweden

The yearly turnover of all betting in Sweden is currently around $4.86 billion with betting on international sites excluded. ATG, which is a large, state-owned company holding the monopoly of all betting at the trotting tracks in Sweden (27 tracks in all), currently holds a market share of 28%. The company has a yearly turnover of $1.42 billion[1]. In 1998, ATG launched betting service on its Web site, InterBet™. The online business amounted to $8 million during the first year in service. Within two years, it arrived at a yearly turnover of $46 million. One immediate effect of the increasing use of InterBet is a decreasing use of the betting service offered at certain public places (i.e., kiosks and small stores) connected to the ATG national network[2]. The different domains of betting, a number of different sports besides trotting, are competing with each other to reach the same customer group (i.e., the public at large). Despite the different foci on sporting events, they all compete with almost the same slogan—the more you know, the greater chance of winning. In addition, in order to avoid creating thresholds for people not that familiar with statistics and rules of the game in question, for example, services are offered that provide customers with ready-to-go bets. These preformulated bets are handled and determined by experts in the field and have gained considerable interest and demand. One immediate effect of these sports knowledge-oriented betting industries is that they also create demand for relevant event information. Naturally, this is essential for customers in order to place bets that are founded on some kind of knowledge base. This need feeds the business of magazines, related Web sites, expert commentary, and statistics. This has resulted in a thick supply of event-related information. In the following subsection, the emerging problems for this domain are outlined.

Three Emerging Problems

Online betting (i.e., InterBet) and improved media coverage (i.e., increased exposure on television) is believed to be contributing to the decrease of spectators attending the trotting tracks in Sweden. As a result, sponsors direct their interests to where exposure is high (i.e., television broadcasts). There is also heavy competition from abroad, nourishing the trend of people not going physically to the track and instead placing bets online. During recent years, foreign competitors have entered the business field of betting in Sweden. The government-controlled monopoly of betting experienced heavy competition from Internet-based companies competing with a wider range of betting services for their customers, as well as with more attractive odds. It is estimated to already take a market share worth $400 million. This is possible due to the fact that the competing companies are based in other countries, which makes the offers as well as the transactions between business and customer legal, which prevents the existing monopoly from interfering. In addition, government influence on the monopoly governs the approach on betting addiction, which includes ethical issues for business. Currently, there is a lawsuit against the monopoly, since foreign companies are not allowed to advertise in Sweden yet. Given these aspects, the monopoly faces heavy competition in the near future, since the government will not allow them to compete by means such as odds. This is based on ethical concerns. This scenario looks very different in other European countries such as the UK, where betting is a more liberated business. The broad approach of providing betting on a number of events is something that foreign actors already have adopted and currently offer customers to bet on almost any event. The emerging problems can be summarized as follows:

- Trotting track managers experience a decrease of attending spectators, and their average age is rising. Improved media coverage and betting online also contribute to this development.
- Foreign Internet-based competitors are gradually taking market shares of the betting business in Sweden.
- The current trend causes decreased sales in sponsoring, entry fees, and other on-site services (e.g., restaurants).

Trotting track managers in Sweden are alarmed about the consistent decrease of the attending audience. This development has been at work for several years and is accelerating. In the next subsection, we provide a summary of the previous research on developing the mobile spectator support system TrottingPal.

The TrottingPal Research Project

The aim of the TrottingPal research project was to design mobile IT support for attending spectators in order to enhance their experience. TrottingPal provides spectators with timely event information, which can be accessed and controlled

Figure 1. Illustration of the three-phase research approach applied in the project to design and develop TrottingPal (for a more thorough description of the approach, see Dahlbom, 1996; Ljungberg, Dahlbom, Fagrell, Bergquist, & Ljungstrand, 1998)

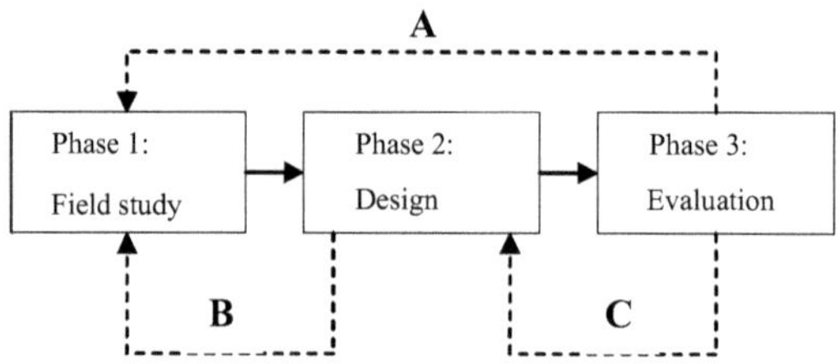

while at the track. The system also allows users to communicate with other TrottingPal users. The system enables spectators to focus less on fixed information sources distributed at the track. The research process applied is illustrated in Figure 1. The main results from each phase will be accounted for briefly in order to illustrate the knowledge that has been transferred from each phase to the next. The collected sets of results are then used as input for the analysis of a potential business possibility. The approach represents the ideal preconditions to conduct the research and its interdependencies between the phases. A field study was done to understand the context and practice of spectators to an extent sufficient for design. Following this, the design was evaluated with target users before implementation (i.e., with a focus group) (see Figure 1, B). The evaluation in this project served several purposes. The main goal was to examine and verify ideas and the system while in authentic use, as opposed to simulations.

The evaluation also provided data to reflect on the quality of the design and also a feedback process on researchers' interpretation of the context and practice (see Figure 1, A, and C).

Field Work and Design

An ethnographically inspired field study was conducted, extending to a total of 50 hours (see Nilsson & Nuldén, 2003, for more details about the study). The main findings (i.e., context and user characteristics) derived from fieldwork were mobility, personalization, and collaboration. These characteristics were found to be most important during the observational study of the spectators. First, spectators are highly mobile during their stay at the track. Mobility stems from gathering information from spatially distributed sources to place well-founded

Figure 2. Screenshot of TrottingPal showing device, interface, and architecture overview

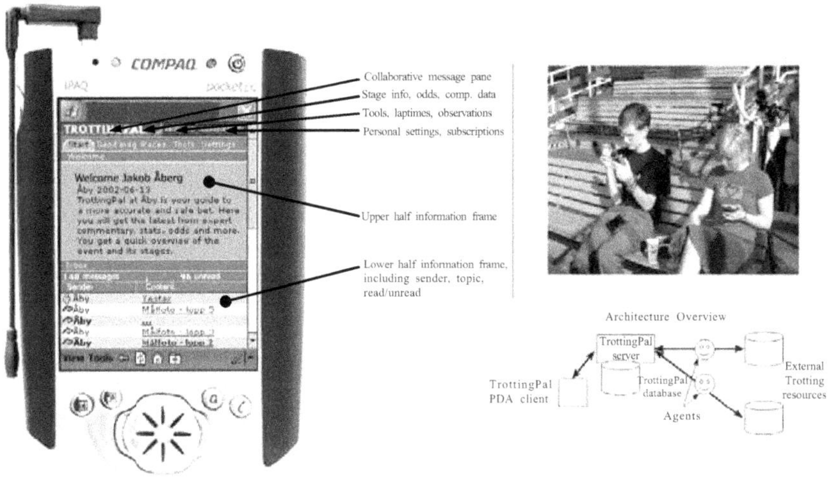

bets. Second, when spectators have varied opinions regarding which information sources to rely on, they personalize their information. Third, spectators jointly collaborate to exchange information to cover a greater set of hints than individually, and to collaboratively interpret and make sense of gathered information. During design, these three main characteristics from fieldwork became gradually more and more intertwined, and the outline of a mobile, interactive multi-channel system began to appear. By using wireless technology (i.e., IEEE 802.11b and PDAs) in an open yet delimited space such as the arena, mobility could be well supported (see Figure 2). To design for personalization, a number of information channels were made available in the mobile system. By applying structure to the collection of information, users could easily personalize and manage their information. Collaborating capabilities were met by integrating simple text message communication.

Still, although the field data showed the importance of a strong link between personalization and collaboration as a preparatory step for placing bets, the project consciously chose not to implement betting.

Evaluation of Technology Use

Effects on how information was gathered, discussed, and treated were identified while using TrottingPal. For instance, briefings that earlier took place in face-to-face situations occurred with the aid of TrottingPal, while being dispersed. In addition, the face-to-face setting was used increasingly for making sense of information. Findings also include changes to how designated spaces on site were used with regard to activity. One example is that places located somewhat away from the track were avoided most of the time and only visited for placing bets. With the support of technology, these spaces now allowed a continuous access of information. Initial results also were collected regarding collaboration among spectators. Interaction among spectators within the system established a common topic, which encouraged users to complement and contribute in order to gather a collective set of information regarding a specific matter (e.g., an anticipated outcome of a race). With the outline of the emerging problems for the betting business and trotting track managers, together with the research background of TrottingPal, the remainder of this chapter will elaborate on the following research question: *What are the benefits of TrottingPal for the stakeholders?*

Electronic Business Meets Empirically Informed Design

This section reviews related contributions concerning mobile commerce and technology, as well as the use of ethnographic techniques to inform design and development of novel services.

Recently, several researchers have proposed a varied range of possible mobile technology use in the context of enabling transactions between businesses and consumers (B2C) (Kannan, Chang & Whinston, 2001; Shih & Shim, 2002; Varshney & Vetter, 2001). Much of these contributions rely on an established or approaching technical infrastructure from which potential services and business models can be drawn. However, few of the proposed services and business models rely on empirical studies. Kannan et al. (2001) provide a set of propositions concerning marketing and marketing research based on a discussion of the characteristics and use of mobile technology. "In order to understand how wireless technology impacts the marketing of goods, services, and content, it is necessary to understand how wireless technology affects some of the aspects of consumer behavior and the choice environment" (Kannan et al., 2001, p. 3). Lehner and Watson (2001) define mobile commerce and outline potential areas of research within the domain of mobile commerce. They define it as "the use of cell phone technology for communications and transactions between an organization and its various stakeholders to improve organizational performance" (Lehner & Watson, 2001, p. 4). In this chapter, the analysis takes its starting point in an empirical case that forms an understanding of user practice and technology use from the previous research on TrottingPal. This research followed a three-phase research approach (Dahlbom, 1996; Ljungberg, Dahlbom, Fagrell, Bergquist & Ljungstrand, 1998; Plowman, Rogers & Ramage, 1995). It also has been referred to as contextual design, where the design of products is based on the designer's understanding of how the customer goes about his or her everyday activities (Beyer & Holtzblatt, 1999). The use of ethnographic field studies to inform design work to shape information technology and its use has gained considerable interest (Button & Dourish, 1996; Button & Harper, 1996; Hammersly & Atkinson, 1995; Hughes, King, Rodden & Andersen, 1994; Plowman, Rogers & Ramage, 1995). To what extent such methods are part of the product work cycle has also been discussed (Bly, 1997). Further, the starting point of examining work-practice for design has been discussed; for instance, when formal descriptions of work serves as input for describing work practice, as opposed to the lived-work (Button & Harper, 1996). This chapter uses the approach of looking at the research results in a broad sense that spans fieldwork, design, and evaluation. To begin with, this helps to achieve a comprehensive base for discerning business potential, looking beyond earlier design delimitations for research that might have relevance for business.

The Potential Business Possibility

This section looks at the previous research effort on TrottingPal in a broad sense in order to identify benefits by relating to several stakeholder needs and goals. In addition, 10 interviews were conducted to collect data on anticipated benefits of TrottingPal for the different stakeholders. The interviewees were people in the sporting event business, mostly within management and sponsoring. The TrottingPal system and its concept were presented to each interviewee. They were asked to rank the top add-on value of this system. The interviews were transcribed and analyzed. The answers were categorized in order to identify important patterns.

The TrottingPal system fulfills the requirements for conducting research and development from an artifact and evaluation point of view. However, it does not accommodate the challenge of discerning potential value for business. This is due to different underlying objectives. The design of TrottingPal focused on incorporating the main characteristics of the spectators from fieldwork (i.e., mobility, personalization, and collaboration). It includes features that would eliminate spectators' forced visits to cashboxes, exchanging them for instant mobile betting, but this was not implemented in the system. The reason was that it would be too expensive and time-consuming to integrate betting functionality with the information tools within TrottingPal, excluding the fact that betting did not inflict any disruption to the research objective described previously. This was done to delimit the design to the main findings. However, betting is an important part of spectator activities. In summary, when the focus shifted toward a

Figure 3. The model to identify potential business value of this particular research project. The scope has been extended with stakeholder goals

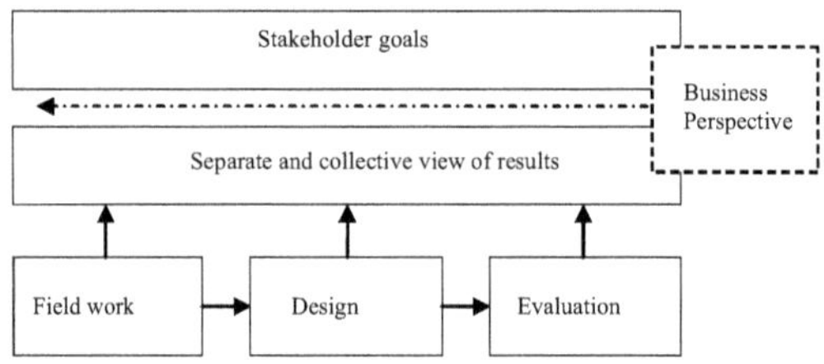

business perspective, examining issues that previously had been ruled out may prove to be important (see Figure 3). The research has addressed support for spectators' main characteristics, as opposed to the business perspective of reaching the fulfillment of stakeholder goals.

The shift in focus from design work to business potential calls for further analysis that examines how the different stakeholders in the value chain as a whole benefit from implementing technology.

Surveying Stakeholder Goals for Business

TrottingPal lacks support for the spectator/stakeholder goal of betting. Further, relating to the evaluation of its use, one constraint is the absence of integration with existing sources of information, which will be a crucial factor in further development due to the importance of information validity. A mismatch, for example, between existing information channels and the content of a new mobile service would instantly make it a total failure (Nilsson & Nuldén, 2003). Since betting functionality is not supported, the link between information and placing bets is nonexistent. As a result, TrottingPal ends up in a situation where it competes with existing information sources and established routines of betting. Still, TrottingPal features functionality that provides far more support for information gathering and collaboration than existing sources. However, the missing link with the betting business threatens the use of the information in TrottingPal when the time comes to place bets. This decreases the potential of TrottingPal to create product demand, as does the poor conditions for integration with the existing value chain.

During design, TrottingPal was shaped to enable the event organizer to communicate information to spectators. This feature enables support for the stakeholder goal of the organizer to direct offers to its customers. Another goal of the organizer is to generate income from advertisement through sponsoring. This part of the business is heavily dependent on the amount of people attending, the rate of exposure. The effect of less attendance not only means decreasing revenue from a smaller audience, but also a weakened position toward sponsors. TrottingPal is focused to add value on site for attending people. It facilitates management of important information at the arena (i.e., observation tools and means for collaboration). By communicating this to spectators, organizers potentially can encourage spectators to attend the track, which can reinforce their relation to other stakeholders such as sponsors. That way, add-on value of new ways to collect and manage information linked to the arena would make the trotting track more attractive for dedicated players. Further, this also can turn around the decreased sales in food at the track mentioned earlier. With an interactive channel open toward spectators, business partners (i.e., restaurants and similar venues) of the trotting track gain a new way of reaching the audience with higher precision in order to improve their offers to spectators and gain competitive advantages. For instance, temporary offers during the evening can instantly be communicated wirelessly to the TrottingPal users. The service provider of betting is ATG facilitated by each trotting track's infrastructure and staff.

At the trotting track, the transactions are done mostly through personal contact with a clerk behind a desk in the designated betting hall. One aspect of this is that the only way to place a bet on site is to pay cash. No credit cards or other means of payment are possible due to privacy issues. Laws against electronically keeping track of identity prevent the organizer and betting service provider from replacing the current solution. TrottingPal could be equipped with an electronic and anonymous value card, similar to prepaid mobile phone subscription. Thus a non-cash solution could aid ATG as well as enable the organizer to reduce substantial cost for cash-box staff. Still, this would require careful redesign and evaluation, since the protection of identity is extremely valued among customers. The overall competitive situation, as described earlier, gives incitement for business stakeholders, such as ATG and trotting track managers, to explore and adopt new means to generate value for customers. This might make way for emerging technologies such as TrottingPal, the introduction of which can be facilitated by young, early adopters. Interactive technology can challenge conventional one-way media support in terms of providing appropriate content at the appropriate time. As one event manager put it, "People easily forget things, to be able to access information when it's needed is very important. To be able to communicate offers to mobile users all the time during the event and at the same time allow for follow-ups is of high value."

It is within reason to assume that information support alone will not suffice to reach success. Instead, the business model could address a multiple set of needs across all stakeholders in the existing value chain. Figure 4 illustrates how mobile

Figure 4. The ecosystem of mobile services in the context of events with regard to the media, sponsors, organizer, and spectators

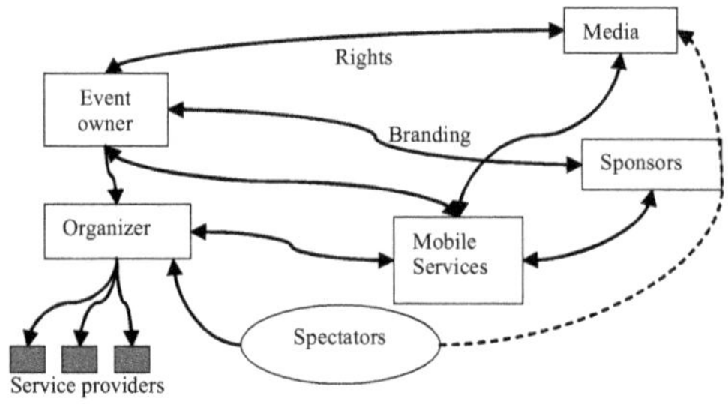

services can be integrated into the existing ecosystem. With an open channel among organizer, sponsors, and spectators, new interactive ways of marketing are enabled from the sponsor's view. For instance, interactive technology between sponsors and end users allows higher precision in follow-ups for sponsors, which makes it easier to survey the effects of different ventures and customer segmentation. A manager of sponsoring claimed, "Customer segmentation is important. Sponsors pay even more for a journalist contact present at the event. Reminders are also potentially important, for instance, about 'small events within the event' when you know they are there!"

Nowadays, the value of sponsoring is measured, for example, by the time of exposure on television (or no follow-up at all).

For instance, sponsoring mediated by TrottingPal enables sponsors to direct timely offers to VIP customers.

Another possibility is to invite segments of customers to on-site venues, which can easily be measured by attendance. This potential, together with betting capabilities, is applicable not only in the context of the trotting track, but from an extended view of crowd-drawing events that include betting. This also might strengthen the position for trotting track managers toward potential sponsors, given the existence of a critical mass of active users. One event organizer talked about the core value of TrottingPal from his point of view:

> *As a tool, this system is of value to the organizer. Especially in terms of increasing the experience of the event, an add-on value for attending people. It would also allow us to extend the event to other venues on site as well as other surrounding events linked to the main event.*

In summary, the implications from surveying different stakeholder goals facilitated by TrottingPal can be described as follows:

- There is much to gain from the implementation of interactive information support on site at the trotting track. New technology, shaped according to the real practice and needs of attending spectators, can attract a younger audience and increase the value of locality and attendance accordingly at the trotting track.
- To introduce novel mobile technology requires clear integration with existing information support to ensure a real sense of information validity.
- To accommodate spectator stakeholder goals, novel information support needs a close integration with betting capabilities. Still, maintaining a high level of privacy is of high importance for novel services of this kind and considered one of the biggest barriers for success.
- By implementing TrottingPal, trotting track managers gain a powerful set of channels that strengthen their position toward sponsors and on-site business partners.
- Multi-channel systems such as TrottingPal can, from a betting perspective, cut costs on cash-box staff and, more importantly, enable new betting services for ATG with higher resolution linked to the course of events at the track.

Discussion

Trotting track managers in Sweden are alarmed by the decreasing amount of attending spectators. It is often compared to the audience situation in the U.S., where few spectators attend the tracks nowadays, and betting is handled remotely online. Still, the development in Sweden is more or less considered inevitable, partly due to the emerging competition from other betting service providers and problems attracting a younger audience. In the future, the competition is likely to focus on increased service for concerned stakeholders, which might find considerable support from the use of mobile IT, as shown in this chapter. By revisiting the research results on TrottingPal, a number of potential ways for add-on value has been identified, spanning organizer, sponsors, betting service provider, and spectators. The core value that TrottingPal can represent in this value chain generally can be referred to as interactivity. This interaction, facilitated seamlessly with mobile technology, maintains and mediates interaction between the stakeholders, as opposed to existing one-way communication. Seamless interaction in the ecosystem of the trotting track enhances and speeds up communication among the stakeholders and renders novel ways to create and measure offers. Ultimately, this may strengthen the locality of the trotting track and provide trotting track managers with new means to attract an audience.

The challenge that design and development of new technology ultimately face is to achieve a change in end-user behavior, to adopt and integrate something new, and to allow it to be cultivated locally and personally. At the same time, it has to fulfill customer needs. This chapter has looked into some aspects that show the complexity in terms of the coming requirements on a design that is to be introduced as a business venture. Research methods such as ethnographically informed design provide us with an understanding of how people go about doing the things they do and why. However, it is important to recognize that somewhat different requirements are added to the scenario from a business perspective. From that point, the design and add-on value needs to fulfill more than user support (i.e., to accommodate an existing value chain). The analytical approach applied in this chapter serves as a starting point to look at the gap between empirically driven design research and requirements from a business perspective. The research is done by looking at an ideal, delimited world so that it becomes manageable to explore in a scientific and systematic way. The business perspective is about incorporating innovation and value in a mature product that is demanded at such a rate that the cost to produce it is less than the money people are willing to pay for it. The problem of launching something created, given a predictable, delimited piece of the world, is that it involves competition from other services and barriers from existing, cultivated artifacts and routines; it also reveals issues, for instance, as to what extent it can accommodate the needs of multiple stakeholders, if one alone does not suffice. This chapter suggests an extension of the perspective of analysis to relate to stakeholder goals. In the case of the trotting track, it is argued that the support for mobility, personalization, and collaboration does not suffice for business potential, viewing it as an information support service. Instead, it is more realistic to regard what TrottingPal addresses as *practice support*, where practice is defined as the activities that spectators go about doing at the trotting track. This is only partly true, since the practice, as such, involves more. For example, mobility, personalization, and collaboration can be described as the preparatory step in order to place well-founded bets, where betting is an important stakeholder goal for spectators. By extending the perspective from spectators to also include the ecosystem as a whole (i.e., other stakeholders at the trotting track), a broader set of benefits has been discerned.

Conclusion

The aim of this chapter has been to examine the ways in which possible business potential can be discerned by relating the results from a design-oriented project to the different stakeholders. Thus, the research question raised was, "What are the benefits of TrottingPal for the stakeholders?"

This chapter brought two main contributions forward. The chapter concludes that the challenge for design-oriented research in the attempt to reach a product stage can find support in extending the perspective and can relate to stakeholder goals. In retrospect, the original research objectives have been achieved from the view of providing IT support for spectator practice, which is supported by the evaluation results. However, although needs of spectators have been identified and designed, the prototype representation does not secure or incorporate potential for entering a value chain. As such, supporting practice is not necessarily the same thing as accommodating stakeholder goals.

The second contribution is that by extending the scope to relate to stakeholder goals, benefits for developed relationships have been identified, mediated by seamless interaction between stakeholders. Interactive technology enables organizers to dynamically update programs and direct offers toward attending spectators. Moreover, it also provides the organizer with an increased value to offer sponsoring parties; namely, that an open channel is available to reach the audience in a new fashion. This potentially strengthens the locality of the track (i.e., the importance and advantage of attending). Generalizing this feature implies a potentially better customer segmentation that could be applied to different types of events (i.e., temporary sporting events). Thus, sponsors could direct special invitations to important customers in real time. Another potential benefit is to cut costs regarding staff handling betting transactions. Given the interactivity of mobile IT, new forms of betting are rendered possible. For instance, when betting transactions can be carried out rapidly between customers and the service provider, betting may achieve a higher resolution. Lately, this development has started to emerge. One example is different sporting events in Sweden where it is possible to place bets while the game is in progress (e.g., which team scores the next or last goal in the match). In the context of the trotting track, this potential might be of great interest for managers, since that type of high precision betting would indeed reinforce locality of the track.

Endnotes

1 For more information, see www.atg.se (in Swedish).

2 ATG owns a nationwide fiber-optic communication network, which is dedicated (i.e., ensuring security, etc.) to running their service at retailers available in public places over the country, such as small stores.

Acknowledgments

I would like to thank the editors: Markku Tinnilä, Timo Saarinen and Anne Tseng, and the reviewers for useful comments.

References

Beyer, H., & Holtzblatt, K. (1999). Contextual design. *Interactions*, *6*(1), 32-42.

Bly, S. (1997). Field work: Is it product work? *Interactions, 4*(1), 25-30.

Button, G., & Dourish, P. (1996). Technomethodology: Paradoxes and possibilities. *Proceedings of the Conference on Computer Human Interaction (CHI),* Vancouver, British Columbia, Canada.

Button, G., & Harper, R. (1996). The relevance of work-practice for design. *Proceedings of the Conference on Computer Supported Cooperative Work (CSCW),* Norwell, Massachusetts, USA.

Dahlbom, B. (1996). The new informatics. *Scandinavian Journal of Information Systems, 8*(2), 29-48.

Hammersly, M., & Atkinson, P. (1995). *Ethnography: Principle in practice.* London: Routledge.

Hughes, J., King, V., Rodden, T., & Andersen, H. (1994). Moving out from the control room: Ethnography in systems design. *Proceedings of Conference on Computer Supported Cooperative Work*, Chapel Hill, North Carolina.

Kannan, P.K., Chang, A-M., & Whinston, A.B. (2001). Wireless commerce: Marketing issues and possibilities. *Proceedings of the 34th Hawaii International Conference on System Sciences,* Maui, Hawaii, USA.

Ljungberg, F., Dahlbom, B., Fagrell, H., Bergquist, M., & Ljungstrand, P. (1998). Innovation of new IT use: Combining approaches and perspectives in R&D projects. *Proceedings of the Fifth Biennial Participatory Design Conference*, Seattle, Washington.

Lyytinen, K., & Yoo, Y. (2002). Research commentary: The next wave of nomadic computing. *Information Systems Research, INFORMS, 13*(4), 377-388.

Nilsson, A. (2004). Using IT to make place in space: Evaluating mobile technology support for sport spectators. *Proceedings of the 12th European Conference on Information Systems,* Turku, Finland.

Nilsson, A., & Nuldén, U. (2003). TrottingPal: Designing for mobility, personalization and collaboration at the trotting track. *Journal of Managing Leisure, 8*, 154-167.

Plowman, L., Rogers, Y., & Ramage, M. (1995). What are workplace studies for? *Proceedings of European Conference on Computer Supported Cooperative Work (ECSCW),* Stockholm, Sweden.

Sarker, S., & Wells, J. (2003). Understanding mobile handheld device use and adoption. *Communications of the ACM, 46*(12), 35-40.

Semrau, M., & Kraiss, A. (2001). Mobile commerce for financial services—Killer applications or dead end? *ACM SIGGROUP Bulletin, 22*(1), 22-25.

Shih, G., & Shim, S. (2002). A service management framework for m-commerce applications. *Mobile Networks and Applications, 7*(3), 199-212.

Urbaczewski, A., Valacich, J., & Jessup, L. (2003). Mobile commerce—Opportunities and challenges. *Communications of the ACM, 46*(12), 31-32.

Varshney, U., & Vetter, R. (2001). A framework for the emerging mobile commerce applications. *Proceedings of the 34th Hawaii International Conference on System Sciences,* Maui, Hawaii, USA.

Section II

Markets, Strategies and Networks: How to Succeed in E-Business

Chapter X

Mobile Games: Emerging Content Business Area

Tommi Pelkonen, Helsinki School of Economics, Finland

Abstract

This chapter analyzes the development trends in a special field within multichannel e-business, digital games designed for mobile devices. It presents frameworks with which to analyze business models, industry positions, and strategic alliances of mobile device game developers, publishers, and telco operators. The key conclusions of the paper are that: (1) game developers should focus more throughly in the creation of excellent and value-adding game titles for consumers; and (2) mobile operators and mobile game pubishers should work actively to create a feasible business environment for market actors and to encourage consumers to consume mobile games. Furthermore, the chapter suggests that the key location in mobile entertainment is shifting gradually from Northern Europe to Asian markets.

Introduction

The mobile channel is emerging as a payment and content consumption platform as well as a business-to-business solution area. Since the late 1990s, one of the key development drivers (outside mobile telephony) within the mobile telecommunications business has been mobile games. The industry comprised of various actors creating, publishing, marketing, and distributing mobile games formulates a very interesting analysis area within multi-channel e-business. As an essential part of the cross-media publishing business, for example, there are opportunities to evaluate a development and value creation pattern within complex industrial networks.

Though growth in the industry is taking place in the early 21st century, mobile computer games[1] have their history in 1980s' portable LCD game series *Games & Watch*[2] from the Japanese Nintendo. These simple miniature consoles introduced a concept of keeping a small source of digitized entertainment in the pocket and constantly available. Mobile gaming (games on mobile phones) emerged with Nokia's launch of the *Snake* game in 1997. In 2003, Nokia introduced its N-Gage game deck and wireless online gaming with N-gage Arena[3]. In 2004, the competition in mobile game markets intensified further with world-leading game titles and with the introduction of wireless-network-connected devices by console manufacturers, Sony, and Nintendo.

This chapter aims to evaluate selected key business trends in the mobile game markets. The key objective of this chapter is to *analyze and discuss* the challenges in the emerging mobile game industry. In addition, this chapter aims to provide insight into the game industry's business logics and value creation processes. In practice, the objectives are met by answering the following research questions:

1. What are the main characteristics and development trends in mobile game markets?
2. How are mobile game market actors interrelated within the industry value creation process?
3. What are the key revenue and cost models within the mobile game development business?

In analyzing the area of games, four different perspectives can be taken—social, individual, technological, and business. This chapter focuses on the game business, highlighting its mobility-related issues. Figure 1 illustrates the analysis focus. Most research related to games focuses on the three other sections[4]. This chapter aims to fill the gap left by the other research initiatives.

This chapter is not a full-scale market analysis or an academic article. A recent EU-funded research project, M-Gain (2004)[5], has published a very thorough analysis of the entire mobile entertainment market. This chapter aims to complement the M-Gain analysis with illustrations of the business markets and revenue models. Furthermore, this report is written from a Northern European perspective of the market. Therefore, generalizations on a global perspective are

Figure 1. Different analysis perspectives to games vs. the chapter focus

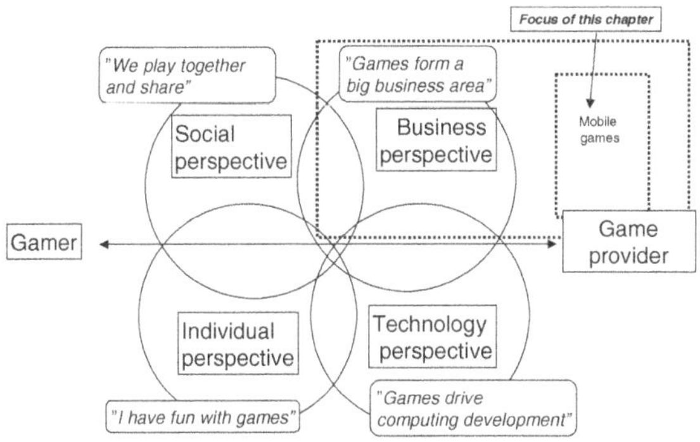

not made. Yet, the business models and industry key findings do have a significant role in the U.S. markets. The analysis is based on the experience of the author in the mobile game markets as well as on several secondary sources. In addition, this chapter is based partly on the author's report published at the European Union founded in the AcTeN Content Market Monitor Series (2004)[6].

Characteristics of Mobile Game Markets

Two Original Business Areas

Emerging mobile game markets are related closely to two existing business areas: (1) mobile telecommunications content business and (2) computer game publishing business. In both of these content creation areas, a similar value creation pattern exists. As in any content business, it has four main operational phases: content creation, content aggregation, content marketing, and content distribution.

The key actors in the markets are game developer companies and game publishers. In the traditional game publishing business, game distributors carry out marketing activities, and sales take place mainly in retail outlets. In the mobile game business, telecom operators and online/mobile portals generate most of the sales and implement to a great extent the marketing activities. In some cases (e.g., N-Gage cartridges), mobile games are also sold in retail stores. Consumers (i.e., gamers) act as the key revenue source for the whole industry, and the money they generate with their game title purchases is distributed to the rest of

Figure 2. Traditional and mobile game value Web

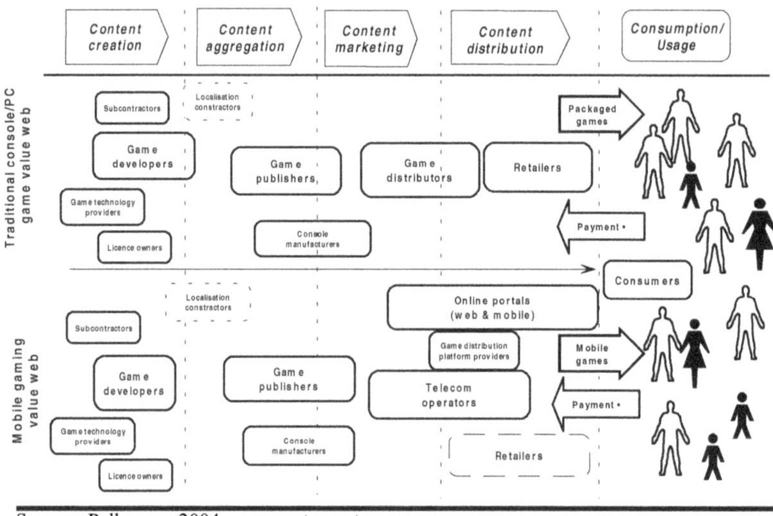

Source: Pelkonen, 2004, www.acten.net

the industry's value chain/Web. Figure 2 presents the game industry value chain and main actor groups in the markets.

The mobile game business is short and cyclical. A new game title is attractive to the market for a limited time, usually selling well for approximately six to 12 months, depending on its geographic distribution range. In addition, the mass markets for mobile game titles are very global.

A company aiming to gain major revenues in the market has to find methods and activities to bring its products to the main markets, such as Japan, the U.S., and Korea. This may become a major obstacle for a small company starting its operations in the mobile game business. The small company needs to find strong distribution partners to help it in its efforts. In addition, a game developer has to find a means to localize its productions to the key markets in the most efficient manner.

Risky Business Area

Game development is a very risky business. Only one out of 10 titles becomes profitable[7]. Yet, the revenues generated by this 10% or less can be very high, and, thus, the potential rewards may balance the business risks. Key elements of the game industry risk are high production costs, timing of launch, and access to distribution networks/chains. Mobile game development has been less capital-

Figure 3. Illustrative earnings cycle for a mobile game title

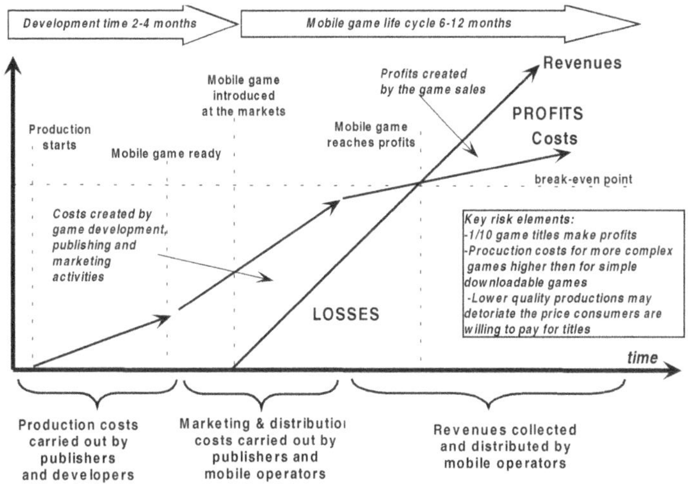

Source: Pelkonen, 2004, www.acten.net

intensive than the traditional console/computer game publishing business. While it takes a minimum of 1 to 3 million euros and 12 to 24 months to develop a console game, it takes only a few hundred thousand euros and two to four months to develop a simple mobile game and four to six months for a more complex mobile game. The mobile game development cycle is illustrated in Figure 3.

Game console manufacturers sell their devices at a deficit. As nearly all device manufacturers operate also as game publishers, they make most of their revenue from game software sales. In mobile games, the console is, in most cases, a mobile phone, and, thus, the case is slightly different. Games are sold as excess service from the mobile operator in order to broaden the phone usage. A mobile phone is still mainly a phone, though having game functionalities. The device subsidy, similar to the game console manufacturer's front-loss, is carried out in most European markets by telecom operators. They provide mobile phones to their customers at a deficit, but tie their customers in to their networks with long-term (24- to 36-month) contracts. The revenues are thus generated later by phoning and data services usage (e.g., game downloads).

Mobile Game Production Process

Mobile game design is an interdisciplinary, collaborative effort. People within the production process work with each other to keep budgets, schedules, and the

Figure 4. Mobile game market competencies

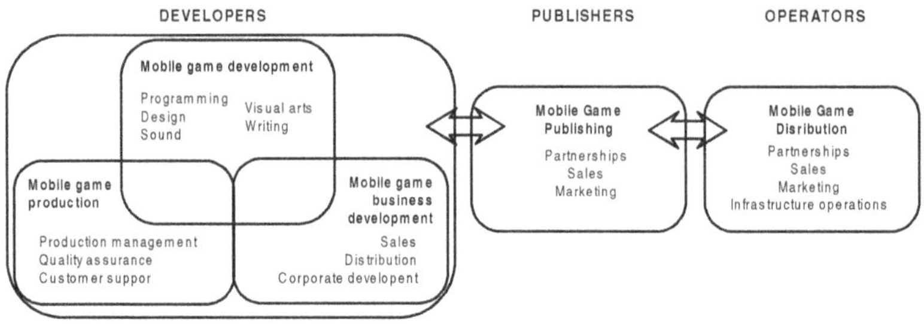

Source: Pelkonen, 2004, www.acten.net

feature and bug lists up-to-date. On the development side, artists and designers collaborate across disciplines to bring the game to life. Finally, production and development teams cooperate with each other in order to keep the project on track and running smoothly[8]. Meanwhile, various kinds of business personnel negotiate on game distribution, project management, sales, and corporate management issues. The simplified operational and competence model is presented in Figure 4.

The more advanced the technology used in game development, the more complex and time- and resource-consuming the production. A simple Snake-like mobile game takes only a few hours for a single programmer to develop. In comparison, a full-scale, advanced mobile console game development may take many developers more than a year to create.

Multiple standards create major challenges for mobile game developers[9]. A game created with one technology solution has to be recreated for another platform. Different screen sizes, varying sound solutions, operating system differences, and memory shortages limit the possibilities for rapid game software transfer from one platform to another. This influences directly the profitability of multi-platform game development. Several development companies have selected only a few technology solutions and game platforms that they support.

Current Market Trends

Growth Business, but Turbulent

The gaming business (including PC, console, and mobile games) was valued at a total of ~USD 25-30bn in 2003[10]. The mobile game business comprises approximately $1 billion or 3% of the total game market[11]. Yet, the growth rates within this area are forecast to be very strong. Analyst predictions on the total future volume of global mobile game business vary from one analyst to another.

Strategy Analytics[12] predicted that the mobile game market would grow to $7 billion by 2008. Informa Media Group[13] has stated that the mobile games segment of the videogames industry would comprise 11.7% by 2006. With no clear agreement on the total amount of the expected turnover, most of the industry players have indicated, nevertheless, that business is growing faster than expected (e.g., German mobile portal Jamba has reported 300,000 Java application downloads per month, out of which 80% are games[14].

U.S. and Japan Lead the Show

The leading traditional computer game markets have been in the U.S. and Japan. Hence, the most significant game industry innovations commonly emerge from these markets. Among the world's leading top 20 game publishing companies, only three have their headquarters in Europe (see Table 1). Yet, in the mobile telecommunications content business, European companies are in a much stronger position. In Europe, the mobile game business has developed rapidly due to the strong support from device manufacturers and the high penetration rate of mobile telephony.

It is expected that when the mobile game business reaches mass markets, the large Japanese and U.S. game developers and publishers will increase their market presence. Actually, this trend is constantly increasing with, for example, Electronic Art's strong entrance into the Nokia N-Gage game cartridge market. The smaller European game development companies will have to cooperate more and more with the game industry giants (e.g., the leading Finnish mobile game developer, Sumea, was acquired in June 2004 by U.S. mobile game startup Digital Chocolate)[15].

Table 1. Top 20 global game developer companies and their turnover in 2002 (European companies in bold)

1. Electronic Arts (USA), $2482 mm	11. Squere Enix (USA), $526 mm
2. Sony Computer (JP), $2180 mm	**12. Unisoft (FRA), $494 mm**
3. Nintendo (JP), $2128 mm	13. THQ (USA), $480 mm
4. Activision (JP), $864 mm	14. Capcom (JP), $407 mm
5. VivenviUniversal Games (USA), $832 mm	15. Bandai (JP), $372 mm
6. Take-Two (USA), $793 mm	16. Namco (JP), $360 mm
7. Atari (FRA), $761 mm	17. Acclaim (USA), $269 mm
8. Konami (JP), $740 mm	18. Koei (JP), $224 mm
9. Microsoft Game Studios (USA), $614 mm	**19. Eidos (UK), $197 mm**
10. Sega (JP), $563 mm	20. Midway Games (USA), $190 mm

Consumers Increasingly Demand Quality

The initial price for a mobile game has been relatively low—~2-7 euros/game. Multiple early adoptors in the game business are testing and buying mobile games at this price. The only exception in game pricing has been the N-Gage cartridges priced at 20-40 euros/title. Yet, these are much broader and more complex games than the compact downloadable games from the operators' networks.

When consumers get used to purchasing mobile games, it is clear that the demand for game quality will increase. Players expect the service provider to guarantee that the value of the purchase matches the price paid for it. If there are a lot of disappointments, the industry may be harmed by a declining reputation. It is important to price games according to their perceived value. Older games should be cheaper than new titles (e.g., launch of the new FIFA Football mobile 2005 would immediately decrease the price of FIFA 2004 game at a mobile portal).

Currently, the most sold game genres[16] among consumers have been mainly action-driven games (i.e., space, shooting, sports, and especially racing). When mobile games increase in popularity more among consumers, other kinds of game solutions will appear. An interesting development in this area is location-based gaming (i.e., games that combine player locations, which are identified by mobile networks or GPS satellites, and actual game play).

Business Models

Key Objective: To Make Profit

As in any other business, the key objective for a game business-focused company is to make a profit. Profits are created simply when companies' revenues exceed the costs. Within the value chain (see Figure 2) for the mobile gaming business, the key revenue source for the industry as whole is the consumers and their willingness to purchase mobile game products. Consumers either pay a flat fee for the game or subscribe to a monthly scheme of payments.

Additional revenue sources consist of various kinds of advertising and sponsorship deals. Advertising can take place either at the place of distribution (i.e., online, mobile, retail) or in the mobile game itself. This kind of sponsorship is similar to sponsorship of television shows and product placement in movies. The game might take place in the preferred location to the brand, or the brand may be brought somehow present to the game play, as it is commonly done in movies[17].

Table 2. Key revenue sources and cost items in the mobile game business

Key Revenue Sources	Key Cost Items
Game Distribution and Marketing Business (20-25% Total Revenues)	
Direct sales to consumers (flat fee) Subscription-based fees (e.g., monthly) Additional feature sales (e.g., extra levels) Game rentals (pay-per-play) Advertising and sponsorships Game merchandise sales Network access and transfer fees (for telecom operators)	Personnel costs Marketing campaigns (planning + media) Infrastructure building and operation costs (distribution platforms, servers, mobile network operation) License fees to publishers Manufacturing fees of merchandise
Publishers (50-60 % of Total Revenues)	
Wholesale game revenues (flat fees/per sales/per usage) Subscription fees from consumers (e.g., to online communities) Advertising and sponsoring (e.g., to include branded content in the game)	Personnel costs License fees to brand owners License fees to console manufacturers Payments to developers
Mobile Game Developers (10-15% of Total Revenues)	
Flat fee payments from publishers Commissions per mobile game usage/sales	Personnel costs Game development software licenses and costs Game development hardware costs License fees to brand owners
Mobile Game Console Manufacturers	
Mobile console sales (Subscription fees to online communities)	Personnel R&D costs Marketing costs Distribution costs

In addition to direct sales and promotion of game sales, various kinds of merchandise products increase the revenues from game titles. These include posters, game character miniatures, clothes, toys, and the like that are related to the game. As in the motion picture industry, the importance of this business area as profit creator can become very important. Hence, it is constantly increasing its role for game marketers and business planners.

Distribution of revenues between different markets is the key issue in understanding the business models within the industry (see Table 2). For example, in 2003[18], mobile game publishers took an average 50-60%, mobile operators 20-25%, licensors 15-20%, and the game developers were left with 10-15% of the total revenues. Advertising and sponsorship deals may naturally change the distribution ratios. Table 2 summarizes the key revenue and cost items for industry actors.

Critical Success Factors

A mobile game developer operation is a very turbulent, rapidly developing, and challenging field of business. To succeed under those conditions, the developer has to focus on the essential—creating excellent game titles. The rest should be left up to the partners. The key business partners for a mobile game developer as well as for game publishers are telecom operators. In the AcTeN Business roundtable discussion[19], Jari Viitanen from TeliaSonera, leading Nordic telecom operator, listed the development areas to which a mobile game developer should pay attention. These are listed in Table 3. As in any partnership, the wish list reveals the need for open attitude and beneficial business models, which benefit all participants of the agreements. In addition, long-term commitment is called for.

From the telecom operator's point of view, mobile games are an excellent item with which to generate revenue. In addition, as content items, they are lucrative and simple to market for consumers. Due to the business rationale created by these facts and, of course, by real customer demand, this area has developed rapidly. Furthermore, dedicated mobile game consoles (i.e., Nokia N-Gage) will increase public interest in mobile games. Yet, as already mentioned, telecom companies should focus more clearly in providing value for their customers. WAP gaming was, in its early years, a major disappointment. Mobile games should not become another area of excessive hype but of marginal revenue. Taking risks may generate a tremendous amount of revenue. Yet, too little risk taking will not generate any revenue for the industry as a whole.

To summarize the business model discussion, there are several critical points that all actors in the mobile game business should consider. These include:

- Partnerships among operators, publishers, and game software developers
- Flexible billing systems that separate games from business content
- Ability of service providers to exploit the networked nature of online games, leveraging the online gaming phenomenon

Table 3. Teleoperators' wish list to mobile game developers' activities

1. Creation of attractive and appealing games for "Young in Mind" segment - Good games for mid- and low-end phones, product roadmap = cooperation tool 2. Not too high prices, preferably teasers for low price; additional levels cost more 3. Flexible approach toward revenue share models (not just selling) 4. New versions for best games (e.g., Lord of the Rings I, II, III) 5. Product availability for different handset models - Testing procedure and handset compatibility (Quality!)	6. Different promotion and delivery models - Own marketing, sales channel cooperation, prepaid 7. Technical and product support and (operator) customer service 8. Pre-review for upcoming product concepts (look and feel) 9. Delivery of complete distribution packets, sales, and promotion support materials 10. Cooperation, active partnering, and change of ideas

Source: Modified from Viitanen, TeliaSonera (2003)

- Ability to host games that consumers can access on a fixed network and upload to their handsets—cross-media leverage and promotions
- An environment that provides games and also enables people to meet other like-minded game players—strong support for online communities

Discussion and Conclusion

Mobile game markets are evolving rapidly and becoming a real business for the industry actors. Mobile gaming is one key mobile content area that generates profits for telecom operators as well as for their publishers and developers. There are multiple reasons for this. First, the distribution technology in mobile networks is capable of delivering mobile games for consumers. Second, more powerful mobile devices have enabled the development of better games and richer game titles. Third, color screens will increase the possibilities of using lucrative elements in mobile games similar to the more advanced computer games. Finally, gaming has become an acceptable form of entertainment within several consumer segments.

The whole entertainment business has been growing constantly in the past few years. Computer games have been one of the key sources for this growth. Within computer gaming, the importance of mobile games is expected to increase in the near future. The mobile channel will be used both to support existing games and to create totally new gaming experiences. The combination of location-based games, multi-player gaming, and powerful mobile devices most likely will enable much richer entertainment experiences for consumers. Though it is a very promising business area, there are still many open issues related to mobile games. These challenges include game transferability from one device to another, game piracy, and intellectual property rights.

Competition for consumers' time is fierce. There are no real guarantees that computer games will become the favorite form of entertainment for all segments. Currently, most game titles are designed for male gamers, and favorite genres are action-driven. If the mobile game industry would like to reach other market segments, new kinds of concepts should be created. Consumers could benefit from playing mobile games to new levels, reaching certain ranking list status, winning in tournaments, and so forth. Combining mobile gaming with television shows (e.g., trivia and quiz shows) offers an interesting opportunity for mobile phone games and applications to become an essential part of the multi-channel content business. It is even possible that mobile phones may evolve as the preferred channel for interactive television (i.e., replacing digital TV set-top-box-based solutions.

The value chain of the mobile game business has evolved into two main business areas—mobile content business and game publishing. Mobile telecom operators play crucial roles for the whole industry earnings logic. Their marketing efforts in providing the games available to their customers determine whether consumers will increase their interest in the area. The game publisher's and game developer's responsibility should be to create game titles that are of high quality and function well in the scattered mobile technology environment.

Mobile game technology standards are still evolving, and it seems their immaturity will still influence the profitability of the mobile game development business. Game developers are struggling with emerging technologies and very often will have to work hard to make their games available for most of the device masses. The mobile phone's main functionality is still telephony. Thus, game features are not the most crucial elements in a consumer's choice for a new mobile device.

At the end of the 1990s, Europe was the leader in mobile telecommunications development. Yet, recent developments in mobile games have shown that the most innovative game concepts seem to be implemented in Asian markets. The challenge from Asia may have a major impact on the European mobile games industry. The computer game business is dominated by the U.S. and Japan. The future will show this to be the case in the mobile game business.

Endnotes

1 Mobile games are computer games accessed through mobile terminals such as handheld game consoles, personal digital assistants (PDAs), and mobile phones. Mobile games can be downloadable or embedded, single-player or multi-player, and interactive or non-interactive.

2 http://www.gameandwatch.com/

3 see www.n-gage.com

4 see e.g. www.gamestudies.org, www.game-research.org or www.gamesconference.org

5 www.m-gain.org

6 see www.acten.net

7 For more detailed analysis of traditional computer game business, see http://www.igda.org/biz/IGDA_2004_BusinessSummit_Proceedings.pdf

8 see Adams (2003), https://www.gamasutra.com/php-bin/store.php?item_id=296& category=9&book

9 One of the key battle areas is carried out at the mobile device operating level: Microsoft-lead Windows Mobile vs. Nokia-lead Symbian Series 60 systems. Nokia is currently in a strong position, but the true mass markets for smartphones are only emerging.

10 http://www.gamespot.com/news/2004/07/30/news_6103791.html

11 author's estimate based on multiple sources

12 http://www.strategyanalytics.com/cgi-bin/greports.cgi?rid=152003030521

13 www.informamedia.com

14 www.nokia.com/nmic2003/pdf/Jamba_Jens-Begemann.pdf

15 www.sumea.fi/press/040624_digitalchocolate.html

16 Game genres are: (1) action, (2) adventure, (3) strategy, and (4) simulation (see http://www.game-research.com/history.asp)

17 Movies and game design do actually have very much in common in production methodology. It is more and more common that movies, PC/console games, Web sites, and mobile games consist of similar elements and promote the same sponsors/advertisers throughout the digital medium.

18 Revenue and cost estimations are based on actual industry experience of the author in the mobile game markets.

19 See www.acten.net - Business Round Table 11 report

References

Adams, E. (2003, Fall). How a game is built and sold. *Game Developer Career Guide*. Retrieved from *https://www.gamasutra.com/php-bin/store.php?item_id=296&category=9&book*

Alcatel. (2003). *Carrier market development: Mobile games*. Retrieved from *http://www.alcatel.com/carriermarketdevelopment/service_sheets/Mobile_Games_sb/Mobile_Games_A4_sb.pdf*

Begemann, J. (2003). Success cases: Java downloads by Jamba. *Proceedings of the Nokia Mobile Internet Conference*, Nice, France.

DFC Intelligence. (2004). Worldwide game market $33.4 billion in 2008. *GameSpot Online*. Retrieved from *http://www.gamespot.com/news/2004/07/30/news_6103791.html*

ESA (Entertainment Software Association). (2003). *Industry sales and economic data*. Retrieved from *http://www.theesa.com/industrysales.html*

Game Research. (2003). *History and genre.* Retrieved from *http://www.game-research.com/history.asp*

Global Information Inc. (2002). *Worldwide market for video game consoles, peripherals, and software is estimated at $28.8 billion in 2001.* Retrieved from *http://www.the-infoshop.com/press/bc8850_en.shtml*

InStat/MDR. (2002). *Wireless gaming market to grow to $2.8 billion worldwide by 2006.* Retrieved from *http://www.instat.com/press.asp?ID=159&sku=IN020418MD*

Kelley, H. (2003). A day in the life of a game developer. *Game Developer Career Guide.* Retrieved from *https://www.gamasutra.com/php-bin/store.php?item_id=296&category=9&book=*

Kuittinen, I. (2003). *Game business. Housemarque.* Presentation at Helsinki School of Economics Information Technology Program, Retrieved from *http://itp.hkkk.fi/files/newsletters/alumninewsletter03.pdf*

Kymäläinen, P. (Ed.). (2004). *Mgain: Mobile entertainment industry and culture.* Helsinki: Helsinki University of Technology. Retrieved from *www.m-gain.org*

Pelkonen, T. (2004). Mobile games markets. *ACTen E-Content Report 3.* Retrieved from *http://www.acten.net/cgi-bin/WebGUI/www/index.pl/mobile_games*

Pelkonen, T., Melakoski, C., & Rissanen T. (2004). Challenges and opportunities in the emerging mobile game markets. *ACTeN Business Round Table 11.* Retrieved from *http://www.acten.net/chris/ BRT_report11_mobile_games.htm*

Strategy Analytics. (2003). *Mobile gaming: Download applications to drive $7 billion market.* Retrieved from *http://www.strategyanalytics.com/cgi-bin/greports.cgi?rid=152003030521*

Viitanen, J.(2003). Ten commandments for mobile game developers. *Proceedings of the ACTen Business Round Table Discussion*, Tampere, Finland.

Wireless Gaming Review. (2003). *Downloadable cellphone games provide gateway to $7 billion opportunity.* Retrieved from *http://wgamer.com/news/shownews-915.html*

Chapter XI

Digital Television and Multi-Channel in Europe

Janne Orava, Helsinki School of Economics, Finland

Mika Perttula, Helsinki School of Economics, Finland

Abstract

This chapter introduces digital television as part of a multi-channel environment in a European context. It argues that digital television in general is in its early stage, that European markets for digital television are fragmented, that development occurs in a country dependently, and that digital television's development pace will have notable effects for the development of a multi-channel communication environment. Furthermore, the authors hope that understanding the development of digital television in a multi-channel context will not only inform researchers of the digital television and multi-channel communication in general, but also will provide some insight into the challenges they are facing on their way to maturity.

Introduction

Today, analog television is the most powerful and most watched medium all over the world. People spend several hours per day watching television. For example, in Europe, TV is viewed an average of three hours and 33 minutes per person daily (Eurodata TV). Its attractiveness is based on both content and media richness. Content can be either informative or entertaining, while media richness makes TV highly audiovisual. This all makes it a very compelling medium both for consumers and for businesses.

Tomorrow, television will be an even more powerful medium due to its increased capacity to broadcast content and its ability to create interactivity with viewers through digitalization. However, digital television today is taking its first steps, and evolution requires the contribution of every actor in the value chain. Development in Europe is highly country-dependent and will be completed in the early 2010s.

Digital television combined with other digital media provides new opportunities for broadcasting and communication. Combining different digital media forms a multi-channel communication environment, which enables seamless content sharing in different channels and the possibility to interact with the viewer. Compared to the current situation, this provides viewers with more comprehensive user experience.

Development of Digital Television in Europe

The history of digital television dates back to the early 1990s, when digitalization of production process and distribution networks started. The next, more visible step for television viewers was the introduction of the first digital channels like Sky Digital and Canal Digital in the late 1990s. Both Sky Digital and Canal Digital are run by proprietary platforms like OpenTV and Mediahighway. After these proprietary platforms, digitalization spread to publicly-led free-to-air models, which are open platforms. One of the most widespread open platforms used especially in Europe is DVB, which is managed by Digital Video Broadcasting Project (DVB Project). Interactive television middleware for DVB is called Multimedia Home Platform (MHP). Digital television can be transmitted via terrestrial, satellite, or cable. One of the new emerging distribution models for television content is Internet Protocol Television (IPTV).

Table 1 is a schedule for development of digital television (DVB) in Europe. Most of the countries have launched digital television, and first national analog switch-offs are scheduled to take place during 2007 in Finland and Italy. After analog switch-off, all broadcasting is digital, and viewers need to have digital receivers in order to watch television.

Table 1. Development of digital television in Europe (DVB)

Country	Legislation in Place	Digital Television Launch	Analog Switch-Off
Austria	2001	2006	2012
Belgium	2002	2004	2005 (start from Flanders)
Denmark	2002	2005	Not decided
Finland	1996	2002	2007
France	2000	2005	2008+
Germany	2002	2003	Started – 2010
Greece	Planning started 2003		
Italy	2001	2004	2007
Ireland	2001	2004	2015
Netherlands	1999	2003	Starts 2004
Portugal	2000	2004	2010
Spain	1998	2000	2012
Sweden	1997	1999	2008
UK	1996	1998	2010

Source: Based on dvb.org

Digital Television Markets in Europe

There are more than 170 million TV households in EU countries, out of which more than 45 million have digital television either via satellite, cable, or terrestrial. According to Strategy Analytics, the number of digital television households in Western Europe is expected to grow to more than 100 million households by the year 2007. The major European digital television markets are in Germany, UK, France, Italy, and Spain. These countries have almost two-thirds of the TV households in Europe.

Table 2. TV households (millions) and digital transmission channel shares per TV households in Europe (2004)

	TV Households	Terrestrial	Cable	Satellite
Austria	3.28	0%	5%	46%
Belgium	4.4	0%	0%	1%
Denmark	2.34	26%	3%	21%
Finland	2.13	14%	0%	1%
France	22.3	0%	4%	40%
Germany	33.4	1%	7%	7%
Greece	3.5	0%	0%	1%
Italy	21.2	0%	1%	24%
Ireland	1.2	1%	14%	25%
Netherlands	6.7	31%	0%	7%
Portugal	3.13	0%	0%	0%
Spain	11.8	1%	0%	25%
Sweden	3.98	4%	2%	5%
United Kingdom	24.8	8%	15%	28%

Source: Various industry sources

Table 3. Digital TV revenues 2000-2006, estimate (US$ billion)

Year	Europe	USA
2000	6.15	5.76
2001	10.69	9.61
2002	17.68	14.31
2003	27.42	21.19
2004	39.53	28.96
2005	54.60	41.91
2006	72.22	60.76

Source: Paul Budde Communication, 2004

Table 4. Revenues from interactive TV applications and services in Western Europe 2002 and 2008 (US$ million)

Consumer Revenues	2002	2008
TV Messaging/E-mail	10	234
TV Betting/Gambling	10	64
Interactive TV games	36	448
Enhanced TV	35	712
Other iTV revenue	1	26
Total	92	1484

Source: Strategy Analytics' Broadband Entertainment Strategies service

There are several estimates and forecasts on how digital TV markets will grow in the near future. Paul Budde Communications estimates that digital TV revenues will grow in Europe from $40 billion U.S. in 2004 to over $72 billion U.S. by 2006 (Table 3).

On the other hand, Strategic Analytics estimates that revenues from interactive TV applications and services will be approximately $1.5 billion U.S. in 2008 in Western Europe (Table 4). This figure doesn't include revenues from pay-TV subscriptions, t-commerce, and interactive advertising.

The digital television market and its development are very country-dependent. In the following, we address examples of digital TV development in the United Kingdom, Finland, and Italy.

- **United Kingdom:** most advanced digital TV markets in Europe
- **Finland:** a pioneer in adopting MHP standard
- **Italy:** adopting digital TV rapidly and several interesting special characters

Table 5. Digital television subscribers in UK/platform (2004)

Platform	Pay TV	Free-to-air
Cable	2 248 166	2 502 451
Satellite	7 085 000	345 000
ADSL		
Terrestrial		3 386 700
Total	13 847 371[1]	

Source: Ofcom, 2004

United Kingdom

The UK was the first to launch digital TV services in Europe in October 1998. Presently, digital TV penetration is over 50% of households totaling more than 12 million homes. Analog switch-off due date is set for 2010, but the government is facing increasing pressure to postpone it.

Until now, satellite transmission and its proprietary platforms have dominated UK markets, but now, the free-to-air terrestrial model (open platform) has become the fastest growing platform. Success of digital TV in the UK has been a learning and reference point for other industry actors across Europe.

Finland

Finland was one of the first countries in Europe to launch digital TV based on DVB and MHP standards (August 27, 2001). Currently (March 2005), almost 600,000 (almost 25%) households receive digital TV broadcasting. The analog switch-off date is August 31, 2007. In order to reach this goal, penetration needs to grow substantially before that. The latest estimate is that in the near future, digital TV receivers will be sold at the rate of 40,000 to 50,000 units per month.

Traditional model for TV broadcasting in Finland is free to air terrestrial and cable transmission. Satellite transmission is marginal and mostly based on the pay-TV model. Digitalization enables new business models for terrestrial and cable transmission, and broadcasters introduced first pay-TV services in the fall of 2003.

Italy

On January 1, 2004, Italy officially launched its terrestrial digital TV. Initially, geographic reach for services will be about 50% of the population, and provincial capitals will be covered during 2005, extending coverage to 70%. The analog

Table 6. Digital TV households in Finland (2004)

Platform	Households
Terrestrial	370.000
Satellite	30.000
Cable	Few thousands

Source: Various industry sources

Table 7. Digital TV households in Italy (2004)

Platform	Households
Terrestrial	700 000
Satellite	5 000 000
Cable	300 000

Source: www.dvb.org

switch-off date is set for 2007, which makes the transition period very challenging. One interesting feature of the Italian digital TV market is the government's 120-150 million euros subsidy for 2004 to ensure the demand of 900,000 MHP-enabled set-top boxes (4.2% of TV households). However, in order to reach government's goal for analog switch-off, sales of set-top boxes should increase dramatically.

Evolution of Digital Television as a Multi-Channel

Evolution in digital television has led to an increased number of TV channels and broadcasters. Besides this phenomenon, broadcasters also utilize other media and distribution channels in their current media offering. These are mobile, Internet, radio, and print media. These communication technologies enable broadcasters to offer viewers constantly new media experiences and, in this way, to build up more long-term profitable relationships. The reasons for this development are several. From an economic perspective, this development opens up potential new revenue streams, provides more cost-effective ways to distribute the same content to various end-user purposes, and to utilize owned media rights most efficiently. Technology no longer presents the biggest barrier for innovations but requires creativity to utilize new technologies most efficiently. Viewer loyalty is also one critical factor for this development, which has led especially to the development of customer participation and interactive services.

Figure 1 describes how digital TV and multi-channel content and services are delivered from a media company to an end user and vice versa.

1. TV content broadcasting is from one to many, and it is the only one-way delivery channel. Digitalization enables new, more advanced services like electronic program guide (EPG), digital teletext, and applications. Utilizing

Figure 1. Digital TV and multi-channel content delivery framework

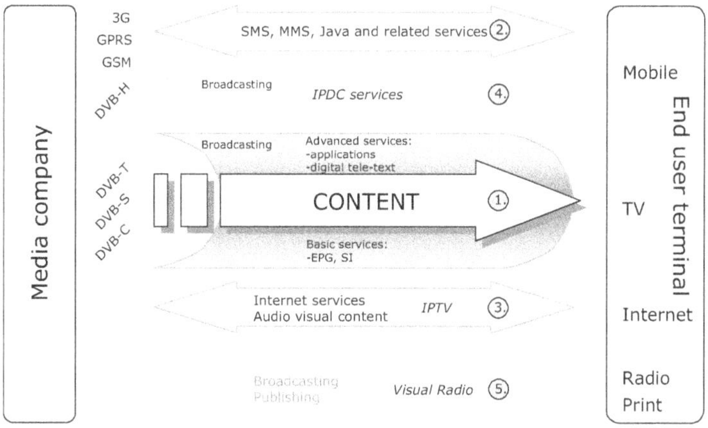

these new services requires advanced set-top boxes (e.g., MHP-enabled set-top box).

2. The mobile channel can be used both as a return channel and as a two-way interactivity enabler between a media company and the end user. The return channel is linked to digital TV content and services, whereas two-way interactivity enables digital TV independent but context-related content and services.
3. Internet access also can be used as a return channel or as a two-way interactivity enabler. Advanced set-top boxes have either modem or ethernet connection enabling Internet access. For both return channel and two-way interactivity, Internet access is used in the same way as the mobile channel. The difference, compared to the mobile channel, is media richer audio and visual content and services because of more advanced terminals (i.e., TV set or computer monitor) and more bandwidth (64-4096 kbps vs. 9.6-384 kbps). In the near future, television content and related services can be delivered using IPTV (Internet Protocol TV) (i.e., all content is delivered using TCP/IP-protocol).
4. The mobile device also can be used as a TV content and services receiver when utilizing IPDC (Internet Protocol Datacast) for broadcasting.[2]
5. More traditional content delivery channels are radio and print, which both are by nature one-way broadcasting and publishing. There have not been any significant advances in print recently. What is new in the radio world is so-called visual radio. Visual radio uses traditional FM radio for broadcasting content, and for services, it utilizes mobile channel for both return

channel and two-way interaction. Visual radio pilot projects started in the summer of 2004.

Case BumtsiBum/MTV3

An example of a television show that utilizes multi-channel opportunities is BumtsiBum on channel MTV3 in Finland. BumtsiBum is a TV program based on an international program format called Lyrics Board. It was first aired in Finland in the fall of 1997 and has been a top viewed program since then. Marco Bjurström hosts BumtsiBum, which has two co-hosts and four changing guests in every show. The main theme is to guess songs based on words and/or pictures from lyrics. This digital TV and multi-channel content delivery framework is shown in Figure 2 to describe a real-life example of how a television channel has utilized multi-channel opportunities in successfully launching a television show.

The BumtsiBum TV program is broadcasted normally. What is new is the interactive Bumtsi game, where viewers can participate and guess songs at the same time as studio guests. After the show, the game provides viewers with score and code that can be sent as SMS to MTV3. Top scores are listed in the digital teletext page after the show. Viewers also can order BumtsiBum ring tones via a mobile channel. The program also has a Web page with background information, picture gallery, forums, and other related content. Also, the media

Figure 2. Digital TV and multi-channel content delivery—case BumtsiBum

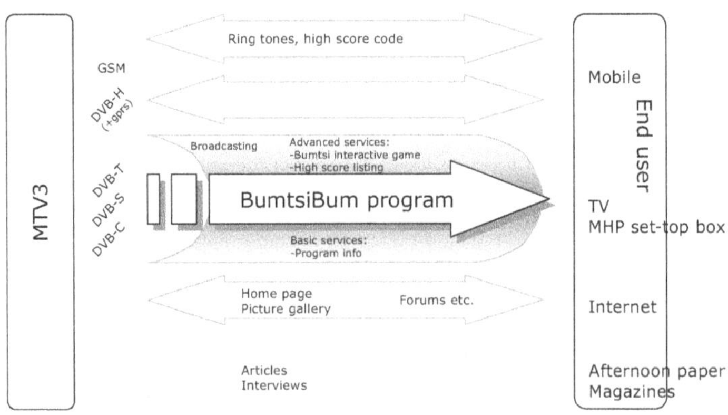

are interested in the program and especially its host Marco. Articles and interviews can be read in afternoon papers and magazines.

Conclusion

After a slow start, markets for digital TV are starting up in Europe (excluding UK). At the European level, markets are fragmented, and development occurs country dependently. Both technology and schedule play significant roles in this development. Matters affecting technological development are standardization, technology choices made by market actors, and general digital TV adoption. Market schedule is affected by the geographical extension of a digital TV network, analog TV switch-off, and general digital TV penetration.

Viewing and user experience for TV content have changed dramatically since the early days of TV. In the beginning, both broadcasting and return channels were analog (e.g., participation in competition by sending a postcard). Then came the first digital return channels, which gave new opportunities for both broadcasters and viewers to experience TV content (e.g., participation in voting by sending SMS). Now, we are in the middle of a transition phase where broadcasting also becomes digital. This development enriches broadcasting by adding new components like advanced value-added services (see Figure 1). The next phase is that broadcasting and return channels will converge. This means one user experience and/or one user interface (e.g., Nokia 7700 and Visual radio).

Digital TV is still developing compared to other media both business- and technology-wise. The true multi-channel world won't be complete before digital TV has reached its maturity and all media have converged. Industry players need to test and pilot various service and channel combinations in order to find the most compelling media offering. When the true multi-channel world is real, TV will become even stronger media, and, with other media, will offer viewers an ultimate media experience.

References

Eurodata TV. (2004). Eurodata TV worldwide presents at MIP TV in Cannes. One television Year in the world-2004 issue. *The analysis of TV viewing and favourite TV programmes*. Retrieved March 5,2004, from *http://*

www.mediametrie.fr/get_doc.php?type=4&id=976& f=d111fb957244ab07b0d079c89292c646

Paul Budde Communications Pty Ltd. (2004). *Europe—Broadcasting—Digital TV.*

Ofcom (2004). *Digital television update—2004 Q3.* Retrieved December 20, 2004, from *http://www.ofcom.org.uk/research/industry_market_research/m_i_index/dtvn/dtn_2004_q3/dtn_2004_q3.pdf*

Strategy Analytics. (2002). *Strategy analytics' broadband device strategies service.*

Endnotes

[1] Adjustment to remove double counting for households equipped to receive digital on more than one set—1,101,280

[2] Although this requires DVB-H receiver in a mobile device (e.g., Nokia 7700 media device with additional Nokia Streamer SU-6 IP Datacast receiver). IPDC broadcasting utilizes mobile access as a return channel.

Chapter XII

Case of Monopolies at Stake: Strategies for Gambling Market

Juha Laine, Helsinki University of Technology, Finland

Jukka Heikkilä, University of Jyväskylä, Finland

Abstract

In this chapter, our purpose is to illustrate the complexity of the outcomes of technological change and the concerns of regulators in the European gambling markets. In 2002, the Finnish Betting and Tipping Company, Veikkaus, estimated, that Finns' Internet gambling decreased its turnover of betting by 10%. Such leakage is a reason why politicians in different states are considering imposing restrictions of online gambling to other parties of the gambling process, such as Internet service providers, banks, and credit card companies. On the other hand, one possible way to caulk the leakage could be to improve the returns of the players, but this has an adverse effect on the social and political objectives of preventing excessive gambling. The recent Gambelli case further indicates that such restrictions might not be considered proportionate and, thus, are against EU law. The diffusion of the Internet also changes the strategies of the industry stakeholders as the interests of potential market entrants, exclusive license holders, regulators, and legislators are now contradictory with the advent of increased gambling possibilities.

Background: Why is Gambling a Monopoly Industry?

Gambling is an industry with high growth potential and payoff for private companies, and it is also a major fiscal factor in the nation states' budgets. The turnover and profit of gambling has increased constantly in recent years, and the gambling industry is expected to continue to grow in the near future. Gambling is correlated with increased wealth (e.g., in the United States, more money is spent on gambling than on groceries) (IGWB, 1998; NGISC, 1999) with estimates ranging between $500-$1,000 billion U.S. per year. There is a general belief that the rest of the world, including Europe, is following in the footsteps of the U.S.. With good prospects for growth and profit, gambling is becoming a major economic factor in the economy. Equally, the importance of revenue from the gambling industry into state financing has increased. This indicates that nation states have an interest in maintaining the monopoly structure of the gambling industry and in banning cross-border supply mainly for fiscal reasons, although it is often argued that state monopolies are necessary to prevent crime and to protect the players.

The increase in gambling is also due to technological convergence, which makes possible the creation of more attractive games (Griffiths, 1999; Järvinen & Sotamaa, 2002). Gambling is increasingly targeted to specific customer segments (e.g., women) via emerging electronic channels.[1]

Gambling can be seen as entertainment, but it also has special negative characteristics compared to other entertainment activities. Some studies explain that gambling is a very tempting means for people to experience excitement, avoid frustration, test limits, or break loose from routines (Kiefer, 1994; Murto & Niemelä, 1993; Zuckerman, 1979). The gambler may behave irrationally, spending money and devoting time to using tricks, inducements, or baits (Jacobs, 1989; Skinner, 1969).

Gambling-related studies generally compare problem situations and pathological gambling with alcohol and drug abuse (Kessler et al., 1994; Shaffer et al., 1997). According to these studies, pathological gambling is just a tiny fraction of other forms of abuse, but it is still considered a significant societal problem, the importance of which is growing (Ilkas & Turja, 2003; NRC, 1999).

First, although the estimated number of problem gamblers is very small (between a few percent and fractions of a percent) among the gambler population, the cognitive addiction to gambling is considered a major problem (NRC, 1999). First of all, this means that dependency on gambling is noticed and understood too late, and only when the gamblers or their relatives face major economic or social problems (Griffiths & Wood, 2000; Nilsson, 2002; NRC, 1999). Second, most studies conclude that it is even more difficult to get rid of a gambling problem than alcohol and drug addictions (NRC, 1999; Prochaska et al., 1992; Thygesen & Hodgins, 2003). Third, problem gambling is increasing among the young, minorities, and low-income classes (NRC, 1999). Gambling addiction can cause social or moral hazards to players and their families. Gamblers can also be prone to criminal activities like money laundering or tax evasion.

These findings and arguments suggest that the social cost of gambling is high. However, recent studies weighing economic and social benefits vs. costs of gambling have shown mixed results. Generally, the social cost of gambling has been considered higher than its benefits. However, this has been challenged by some of the studies (Dickerson et al., 1995; Thompson et al, 1995), in which researchers have found gambling profitable, even when counting all related direct and indirect costs. Because of this controversy, in most societies, gambling is considered if not a major problem at least an industry worth controlling (GRB, 2002; Griffiths & Wood, 2000; NGISC, 1999). The control policies have their critics; it is argued that limiting gambling is welfare diminishing (in an economic sense), it restricts the freedom to use one's wealth as wished, and limitations waste huge opportunities of business in the long run.

The general explanation to the soaring problem of gambling appears to be increasing supply. Therefore, because of its special characteristics, providers of gambling services normally are required to have a license from the nation state where the gambling services are marketed. Perhaps for fiscal reasons, it is common—especially in Europe—for exclusive licenses to be granted to state monopolies. Remote supply of gambling services (i.e., gambling across national boundaries) has also been mostly banned. Until now, remote gambling has been marginal also in reality, because the gamblers have had little information on the overseas possibilities. It has been burdensome and expensive to find out the supply and suitable means of payment. Now, the situation is changing, because remote gambling has become common; for example, Internet gambling has quadrupled in a couple of years in a row among Internet users over 15 years of age (GallupWeb, 2002).

Online vs. Internet Gambling

In this chapter, gambling means that the player participates in a gaming activity with money, the odds have a significant role in the outcome, and the prize is money. There are two basic types of gambling. In games of chance, the outcome is determined solely on the basis of the odds, and the player cannot influence the outcome. Games of chance include, for example, lotto, roulette, and slot machines. In games of skill and chance, the player can affect the outcome with his or her skills and knowledge. Games of chance and skill include games such as betting. The gambler cannot be expected to have an effect on the outcome of the betting, but a well-informed better is likely to gain more than a less-informed better.

Online gambling is most often considered equivalent to Internet gambling, but it is worth noting that digital convergence (Heikkilä & Laine, 2001) means that, in the future, gambling will be available on other types of technical infrastructures like e-mail, phone, fax, SMS, or interactive DVB.

Online gambling is further divided into two types:

1. Games where the network is a channel for a gambler to participate in, but the actual lottery, or determination of the outcome, takes place outside the network. Examples are football pools, betting, and lotto, in which a gambler can participate either with a retailer or over the network. Participation over the network does not change the original character of the game.
2. Games in which a gambler participates only via networks, and the lottery is conducted online. Examples of such online gambling are Internet slot machines or roulette. The games can be substitutes for existing physical games (e.g., virtual slot machine) or completely new types of games.

Regulation of Online Gambling

In the world of the Internet, the regulation of gambling has not been totally effective. There is unlimited information on foreign gambling facilities, guidance on how to participate, and common European currency. Access to universal payment and to user-friendly virtual gambling facilities is diffusing rapidly. All stages of gambling—registration, payments, gambling itself, and cashing-in the prizes—can be done quickly and at a low cost over the Internet.

In online gambling, the gambling operator (i.e., with a license from its nation state) can be outside the direct jurisdiction of gambling regulators from other nation states, even though its gambling services would be available via the Internet in another nation state. This challenges the exclusive national licensing system. In practice, it is often difficult for a regulator to intervene in the foreign online supply of gambling services. There are several ways, however, that it can attempt to limit the supply. First, the regulatory authority could try to intervene indirectly, by trying to convince the foreign gambling provider to cease transnational marketing on a voluntary basis. Second, the regulators could cooperate in order to ask the remote regulator to ban the marketing to other states via the Internet. Third, the nation states could conclude bilateral treaties to end the provision of gambling between the states.

In the United States, there has been a strong policy of banning cross-border Internet gambling at federal and state levels. The 1961 Wire Act states clear penalties with regard to interstate and foreign betting by using wire communications.[2] There is also pending federal legislation called the Unlawful Internet Gambling Funding Prohibition Act that would make it unlawful for banks, credit card companies, and other financial firms to knowingly transfer money to Internet gambling sites. It would also make it a federal crime to operate unlawful Internet gambling (The Library of Congress, 2004).

But even the U.S. has met with legal challenges when Antiqua and Barbua (June 13, 2003) brought a claim to the WTO Dispute Settlement Body that measures taken against the cross-border supply of online gambling services in the United States are inconsistent with the United States' commitments and obligations under the General Agreement on Trade in Services (GATS) with respect to the cross-border supply of gambling and betting services (WTO 2003).

At the EU level, the regulatory situation is more controversial than in the U.S.. The European Court of Justice has constantly confirmed that national restrictions on cross-border supply of gambling services contradict the freedom to provide services under the Treaty. The mutual recognition principle guarantees free movement of services even without the need to harmonize member states' national legislations. Services that are lawfully produced in one member state cannot in principle be banned from another member state. Possible bans or restrictions must be based on overriding general interest such as crime prevention or consumer protection and not on fiscal objectives. The European Court of Justice has the ultimate right to decide whether a national ban or restriction is in line with internal market principles.

Online gambling services are information society services (ISS) within the definition of EU legislation. Information Society Services means any service normally provided for remuneration, at a distance, by electronic means, and at the individual request of a recipient of services. The Electronic Commerce

Directive (2000/31/EC) includes, among other things, the internal market principles for ISSs. The principal rule is that the national provisions of the country where the service provider is established are applied (i.e., the home country principle). Member states may restrict the freedom to provide ISSs from another member state only for certain reasons specified in the Directive (i.e., public policy, public health, public security, and consumer protection). The directive is however not applied to gambling services. This means that the Electronic Commerce Directive does not free up cross-border trade of online gambling services. Member states however, are obliged to notify the commission of all pending legislations dealing with ISSs, including gambling services.

Meanwhile, legal uncertainty provides possibilities for opportunistic behavior. Last spring, the British government legalized Internet sports betting, and it looks like online casinos are next in line to get official support. Operators will have to be registered as British companies, locate their Web servers in Britain, and use UK Web addresses for casino sites. However, British gambling operators have been tapping the uncertain legal situation by offering cross-border online gambling services to other member states. The British government is not going to set up limitations for cross-border supply of online gambling services (Department of Culture Media and Sport, 2004).

The Jurisprudence of the European Court of Justice

The European Court of Justice (ECJ) has been prone to protect national gambling monopolies at the expense of internal market freedoms. However, the case of Gambelli suggests that it might not be possible to protect monopolies effectively in the borderless digital environment without violating the principles of internal market freedoms. The legal situation is far from clear, as shown in the following case examples.

In the three key judgments—Schindler, Läärä, and Zenatti—the ECJ laid down guidelines by which states could limit or even ban the cross-border supply of gambling services within the EU. Possible restrictions must be based on overriding reasons such as crime prevention and consumer protection and not on fiscal reasons.[3] In the more recent case of Gambelli, the ECJ reconfirmed that public interest considerations may justify limitations on the free movement of services, but only if the restrictions imposed are not disproportionate to the objectives to be achieved. The court also emphasized that a member state cannot invoke public order concerns if the member state incites and encourages consumers to participate in such games. The proportionality principle means that the restrictions must be necessary in order to achieve the justified objective, and that they do not go beyond what is needed to meet the objective.[4] Since the Gambelli decision, the European Commission has partially abandoned its former standstill position on national gambling monopolies and responded with legislations that are in obvious conflict with the proportionality principle of the treaty.

Recent cases in Denmark and Greece illustrate the European Commission's view on what constitutes violations of the proportionality principle. In March 2004, the Commission warned Denmark that it must justify its restrictions on non-Danish bookmakers, or it will face a challenge in the European Court of Justice. The Commission's request, which took the form of a letter of formal notice, formed the first stage in its infringement procedure under Article 226 of the EC Treaty: "Danish law restricts in particular the provision of sports betting services. The commission intends to verify the compatibility of the ban in question with the provisions of the EC Treaty."

The EU Commission also recently ruled that Greek Law Number 3037, which explicitly forbids electronic games with electronic mechanisms and software from public and private places, is not compatible with the provisions of the EC Treaty. The law was introduced in an attempt to stamp out illegal gambling with offenders facing fines of 5,000 to 75,000 euros and imprisonment of one to 12 months. The European Commission has demanded that Greece lift the ban; if it fails to do so, it will also find itself in front of the Court.

Conclusion

This case illustrates the need for government regulators to adapt to industry changes spurred by the emergence of new technologies. It shows that if legal institutions have to intervene on the development of a market, then authorities and politicians have to finally start to rethink their regulatory position on current gambling license holders.

Since the Gambelli decision, the European Commission has partially abandoned its former standstill position on national gambling monopolies. The Commission also has reacted against Greece and Danish legislations, which obviously contradict the proportionality principle of the Treaty. Furthermore, on May 3, 2004, the Commission gave a proposal for a directive on services in the internal market (COM 2004 2 Final/3). Gambling services are finally included in the proposal so that during a transition period, the Commission will assess the possibility of presenting proposals for legal harmonization. In the meantime, the Commission obviously will hold the position that possible restrictions must be based on real public interest objectives (e.g., prevention of crime and protection of players, which means that restricting the service provisioning must be in line with the proportionality principle).

In accordance with EU law, national monopolies cannot be justified purely by fiscal reasons, no matter how noble the purposes of the use of the funds are. National legislators should take this seriously and restrict incoming cross-border online supply of gambling services only for the valid legal reasons listed previously. Legislators also should pay special attention to the use of overly addictive features of the games provisioning and restrict overall marketing and supply of gambling services.

This means that few member states would benefit from trying to object to the initiatives of the Commission, which aims to harmonize services legislation within the gambling sector. In the online environment, the effective protection of monopolies does not seem to be possible without putting sanctions on Internet access operators and payment service operators, or even criminalizing gaming activities. Such sanctions are politically challenging and most likely in contradiction with the proportionality principle. The Gambelli decision indicates that the proportionality principle is to be taken seriously in online gambling services in the future. It is also possible that in forthcoming cases, the European Court of Justice will evaluate the proportionality by itself instead of letting the national courts decide.

As a consequence, the national gambling monopolies are in a problematic situation: despite their monopoly position, they have to react to cross-border online competition. For instance, Finnish Veikkaus in February 2004 was obliged to raise its payback percentage of sports betting to international levels. However, uncontrolled price competition or competition by developing more addictive games might not be in line with the desired objective to protect players.

In our view, this is the right time to try to introduce a common European framework on gambling throughout the internal market. This would not mean an unlimited supply of gambling services within the EU, but the creation of a common regulated but competitive market for gambling services, including transnational gambling. Rather, this would mean that certain gambling activities, such as lotto, could be left totally outside the free market. In other words, lotto does not need to be heavily regulated, if it is left in the hands of state monopolies. This community framework should not only establish the principles for the provision of gross-border e-gambling services and harmonize consumer protection, but it would also give member states a certain degree of flexibility to adopt tailored national or culture-dependent measures, in compliance with the European framework.

References

Case C-275/92. Schindler. Judgement of the Court of March 24, 1994.

Case C-124/97. Läärä. Judgement of the Court of September 21, 1999.

Case C-67/98. Zenatti. Judgment of the Court of September 29, 1999.

Case C-243/01. Gambelli, et al. Judgment of the Court of November 6, 2003.

Case C-42/02. Lindman. Judgment of the Court (Fifth Chamber) of November 13, 2003.

Cividino, A. (2002). Why women gamble online? WINNEROnline.com (online serial), Retrieved 15.12.2004 from *www.winneronline.com/articles/january2002/women_online.htm*

COM(2003) 702. First report on the application of directive 2000/31/EC of the European Parliament and of the Council of 8 June 2000 on certain legal aspects of information society services, in particular electronic commerce, in the Internal market. The European Economic And Social Committee, Brussels.

Department of Culture Media and Sport. (2004). Gambling Bill. Retrieved from *http://www.culture.gov.uk/gambling_and_racing/gambling_bill/default.htm*

Dickerson, M., et al. (1996). *An examination of the socioeconomic effects of gambling on individuals, families, and the community including research into the costs of problem gambling to New South Wales.* Sydney, Australian Institute for Gambling Research, University of Western Sydney

Directive 2000/31/EC of the European Parliament and of the Council of 8 June 2000on certain legal aspects of information society services, in particular electronic commerce, in the internal market (Electronic Commerce Directive),

GRB Gambling Review Body. (2002). A safe bet for success. *UK Department for Culture, Media and Sport, 41*, 46.

Griffiths, M. (1999). Gambling technologies: Prospects for problem gambling. *Journal of Gambling Studies, 12*(4), 471-473.

Griffiths, M., & Wood, L. (2000). *Lottery gambling and addiction: An overview of European research.* Nottingham: Nottingham Trent University.

Heikkilä, J., Laine, J., & Salo, J. (2002). Rajat ylittävien eärahapelien sääntely. *Defensor Legis N:o 6. 2002*, 1018-1043. (In Finnish.)

IGWB International Gaming and Wagering Business. (1998). The United States gross annual wager. *International Gaming and Wagering Business, 19*(8), 3.

Ilkas, H., & Turja, T. (2003). Rahapelitutkimus, Sosiaali- ja terveysministeriö, 15.5.2003. *Taloustutkimus Oy T-5190 HI/Ttu.* (In Finnish). Retrieved December 15, 2004 from *http://www.stm.fi verkkosivuilta*

Järvinen, A., & Sotamaa, O. (2002). Pena: Rahapelaamisen haasteet digitaalisessa mediassa, Tampereen yliopiston Hypermedia laboratorion verkkojulkaisuja I. (In Finnish). Retrieved December 15, 2004 from *http://tampub.uta.fi/tup/951-44-5411-1.pdf*

Kessler, R.C., et al. (1994). Lifetime and 12-month prevalence of DSM-III-R psychiatric disorders in the United States. *Archives of General Psychiatry, 51*(1), 8-19.

Kiefer, P. (1994). *Sensation seeking hos Normala Spelare.* Lunds Universitet. (In. Swedish.)

The Library of Congress. (2004). Unlawful Internet gambling funding prohibition act. HR 2143 RFS. Retrieved from *http://thomas.loc.gov*

Murto, L., & Niemelä, J. (1993). Kun on pakko pelata: Riippuvuus, persoonallisuuden häiriö, kohtuuton harrastus vai eettinen ongelma? Tutkimus suomalaisen ongelmapelaajan profiilista, Helsinki 1993. (In Finnish.)

NGISC. (1999). National gambling impact study commission final report. Washington D.C.: NGISC. Retrieved 15.12.2004 from *http://www.casino-gambling-reports.com/GamblingStudy/*

Nilsson, T. (2002). Jakten på Jackpot. Spelinstitutet. (In Swedish.)

NRC. (1999). *Pathological gambling: A critical review.* Committee on the Social and Economic Impact of Pathological Gambling, National Research Council, USA..

Prochaska, J.O., DiClemente, C.C., & Norcross, J.C. (1992). In search of how people change: Applications to addictive behaviors. *American Psychologist, 47*, 1102–1114.

Proposal for a Directive of the European Parliament and of the Council of 13 January 2004 on services in the internal market. Com 2004(2) Final.

Shaffer, H.J., Hall, M.N., & Bilt, J.V. (1997). *Estimating the prevalence of disordered gambling behavior in the United States and Canada: A meta-analysis.* Cambridge: Harvard Medical School Division of Addictions.

Skinner, A. (1969). *Contingencies of reinforcement: A theoretical analysis.* Engelwood Cliffs, NJ: Prentice-Hall, Inc.

Thompson, W.N., Gazel R., & Rickman, D. (1995). The economic impact of Native American gaming in Wisconsin. *Wisconsin Policy Research Institute, 8*(3), 47.

Thygesen, K.L., & Hodgins, D.C. (2003). Quitting again: Motivations and strategies for terminating gambling relapses. *The Electronic Journal of Gambling, 9.*

Volberg, R.A. (2003). Has there been a feminization of gambling and problem gambling in the United States? *The Electronic Journal of Gambling, 8.*

WTO. (2003). United States—Measures affecting the cross-border supply of gambling and betting services—Request for consultations by Antigua and Barbuda. Retrieved from *http://www.wto.org*

Zuckerman, M. (1979). *Sensation seeking: Beyond the optimal level of arousal.* New York: Wiley & Sons.

Endnotes

1 According to Cividino (2002) and Volberg (2003), "Women represent a rapidly expanding segment of the online gambling population and there are a growing number of specialized Web sites for women gamblers. Online gambling offers excitement and escape but also local availability, flexible hours, a low price of participation and physical and emotional safety—features especially appealing to women."

2 "Whoever is engaged in the business of betting or wagering knowingly uses a wire communication facility for the transmission in interstate or foreign commerce of bets or wagers or information assisting in the placing of bets or wagers on any sporting event or contest, or for the transmission of a wire communication, which entitles the recipient to receive money or credit as a result of bets or wagers, or for information assisting in the placing of bets or wagers, shall be fined under this title or imprisoned not more than two years, or both."

3 Already in the Schindler case, ECJ confirmed that lotteries are services within the meaning of the Treaty. But the Schindler judgment states that, given their peculiar nature, lotteries from other EEC member states can be restricted, even prohibited by a member state, provided that those restrictions are not discriminatory on the basis of nationality. The Schindler case lists the acceptable objectives (e.g., prevention of crime, protection of players and their families, moral and religious issues, and even fiscal reasons), which can justify restrictions on the free movement of services.

These reasons have been confirmed in later cases such as Läärä, Zenatti, Gambelli, and Lindman. In the Läärä and Zenatti cases, the ECJ stated that the considerations set out in Schindler were applicable to other forms of gambling as well; namely, the provision of slot machines and sports betting.

4 Gambelli is also the only case that deals with Internet gambling. This may be the reason for the court emphasizing the "proportionality principle". However, the court did not make the final evaluation itself but sent the case back to the national court and asked it to determine whether the restrictions and criminal penalties were in line with the proportionality principle.

Chapter XIII

Serving Customers in a Hybrid World: Multi-Channel Strategies in Retailing

Claas Müller-Lankenau, University of Münster, Germany

Kai Wehmeyer, University of Münster, Germany

Stefan Klein, University College Dublin, Ireland

Abstract

In contrast to a widely held belief that online and offline channels of click and mortar companies should be tightly integrated, this chapter proposes a taxonomy of four multi-channel strategies that show varying degrees of integration and mutual support of online and offline activities. Each of the four basic strategies is illustrated by an example from the grocery retail industry in order to underscore the strategic choice companies have even within an industry. In a second step, we have developed a set of indices that we use to analyze multi-channel retailers' Web sites. The indices are used to classify and discuss the four case companies' implementations of multi-channel strategies in the Web channel.

Introduction

After the disappointing performance of many Internet pure players, click and mortar channel systems have raised new hopes to achieve profitable e-commerce activities. The Internet and, to an increasing extent, mobile technologies inspire visions of completely integrated online and offline channels that support customer channel switching within and across transaction phases (Rheault & Sheridan, 2002; Steinfield, Bouwman & Adelaar, 2002). Consultants claim that multi-channel customers are more loyal and two to four times as profitable as single channel customers (OC&C, 2002). However, the integration of online channels into an already existing, often complex distribution system confronts managers with a number of challenges (Gallaugher, 2002; Webb, 2002). To ensure economic viability and customer acceptance, retailers face the challenge of defining a multi-channel strategy that creates benefits that outweigh increased costs for both the company and its customers (Anderson, Day & Rangan, 1997; Gadde & Ford, 2000). Next to large, up-front investments for competitive Web sites for online transactions and customer support (Schoenbachler & Gordon, 2002), a multi-channel strategy incurs further costs through channel integration and coordination of a complex system of interdependent channels (Malone & Crowston, 1994; Peterson, Balasubramanian & Bronnenberg, 1997). As for the revenue side, however, there is no obvious reason why the Internet or any Internet-based service will in itself increase consumer spending in a multi-channel system (Peterson et al., 1997). The Internet might even cause consumers to spend less (Deighton, 1997), and a lower service level combined with price reductions can lead to a higher perceived value (Gadde & Ford, 2000). Therefore, depending on a company's situation and corporate strategy, an extensive integration of channels is not necessarily the most promising multi-channel strategy.

The purpose of this chapter is to further understand different multi-channel strategies and their implementation, especially in the Web channel. We develop a framework for multi-channel strategies (in the sense of click and mortar strategies) that is based on channel integration and channel domination as classification criteria. It highlights alternative roles and relative strategic positions of online and offline channels, respectively (Earl, 2000), and is applied to four cases from the grocery retail industry.

In order to support the application of the framework and to generate and explore research propositions, we have developed a set of items and indices to analyze multi-channel retailers' Web sites and, thus, the implementations of multi-channel strategies in the Web channel. We apply this approach to the four case companies' Web presences.

While our work can support business practitioners in strategically positioning or analyzing multi-channel strategies and their implementations, we also demonstrate how the framework can be used in further research activities to generate and discuss research propositions. The propositions concern the role of online customer retention measures and the intensity of online communication and information in the context of the different multi-channel strategies. We conclude with a discussion and an outlook for further research as well as management issues.

Multi-Channel Strategies

Throughout this chapter, the term *multi-channel strategy* is used synonymously with strategies of combining physical and virtual distribution channels. More precisely, we focus on the combination of physical retail chains and an Internet presence.

Literature Review on Classifications of Multi-Channel Strategies

The framework developed in this chapter gives an overview of strategic choices regarding the relative roles of and relationship between online and offline channels in multi-channel systems, with domination and integration as classification criteria. It is an extension and critique of existing frameworks, some of which offer complementary views on multi-channel strategies, while all take different perspectives. Steinfield et al. (2001) (following Venkatesh, 1999) describe six Internet strategies classified into two groups: synergy oriented strategies and strategies not exploiting complementarities. As the term *Internet strategies* indicates, the classification is not limited to multi-channel strategies. As for the synergy-oriented strategies, the classification highlights the roles of online and offline channels in initial channel design (online channel designed like offline channel vs. remaking the offline channel to better support online processes) rather than the roles and relationships of offline and online channels in established channel systems. Wirtz (2001) introduced a framework that can be used to analyze the extent to which a company relies on IT in both its distribution (customer interface) and value creation (back office) activities. Gulati and Garino's (2000) framework of multi-channel strategies focuses on the degree of integration and separation of brand, management, operations, and equity in multi-channel systems. Doolin, McQueen, and Watton's (2003) framework highlights the degree of business transformation caused by the introduction of multi-channel strategies. The framework has been designed for Internet strategies of established retailers and includes the support of traditional channels through an Internet channel but does not differentiate between alternative roles the online channel can play in a multi-channel system.

A Framework for Multi-Channel Strategies

Multi-channel strategies can fundamentally differ regarding the relationship and roles of online and offline channels. In cases in which online and offline channels are used as full-fledged, equally promoted distribution channels, different degrees of channel interdependence can be found. In some cases, online and offline channels serve markets independently. In others, online and offline channels are integrated and mutually support each other. Customers are encouraged to switch between channels in both directions.

Further examples show that distribution strategies can emphasize either the online or the offline channel. The minor channel then plays mainly a supporting role by guiding customers to the preferred channel. Even if a supporting channel enables complete transactions, customers generally will be encouraged to switch to the preferred distribution channel. Figure 1 gives an overview of the four basic multi-channel strategies.

Figure 1. Framework of multi-channel strategies

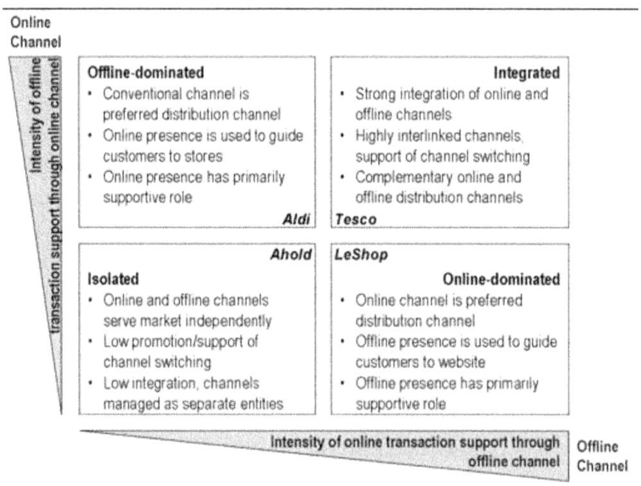

Offline-Dominated Strategy

In an offline-dominated strategy, the online channel plays a supporting role and is mainly used to increase allocation efficiency of the offline channel. Content of the Web site often is limited to (partial) information on in-store offerings and store locations and limited additional services such as after sales information. Possible motivations for pursuing this strategy are a sophisticated distribution system optimized for providing goods to a network of shops; strategic differentiation by means of in-store customer advisory services, which cannot be offered on the Web; or contractual obligations.

Isolation Strategy

An isolation strategy is pursued when online and offline channels are managed as separate and independent entities, possibly operating under different brand names. Neither communication activities, incentives, nor explicit links support or encourage customers to switch between channels. Possible factors supporting a choice of this strategy are the avoidance of channel conflict; targeting different customer groups with channel-specific pricing schemes or serving different geographical regions with online and offline channels.

Online-Dominated Strategy

In online-dominated strategies, the offline channel is configured to guide customers to a corporate Web site or similar online offerings. Communication and promotion activities in the offline realm and possibly incentives such as lower prices or a wider range of products offered online aim to strengthen the online channel. This strategy could be used when trying to bypass intermediaries or to shift sales volume from a cost-intensive offline channel to the cheaper online channel. In an analogy to offline-dominated strategies, previously online-only retailers also can try to gain access to offline channel infrastructure in order to increase their online profitability or sales volume.

Integration Strategy

Integration strategies view all channels as complementary components of a multi-channel system that aims to provide a high level of convenience to customers (e.g., through supporting channel hopping in and between transaction phases). This form of differentiation strategy might help to charge premium prices but also incurs extra costs. In contrast to strategies where one channel dominates, it can be expected that restrictions of product or service offerings in one channel are due rather to technological restrictions than to strategic decisions favoring one specific channel.

Multi-Channel Strategies in the Grocery Retail Industry

Marketing channels provide access to markets and customer segments. They are a core element of business strategy (Coughlan, Anderson, Stern & El-Ansary, 2001). The wide range of functions, roles, and players involved represents a considerable challenge to management (Coughlan et al., 2001). While multi-channel strategies differ due to varying strategic calculi, they have to be analyzed in the context of relevant product, market, and industry characteristics. We chose the grocery retail industry as an example. Based on a brief overview of channel and product characteristics and Internet impact on the grocery retail industry,[1] we will use four cases to illustrate the different multi-channel strategies.

Channel and Product Characteristics

The level of concentration in the grocery retail industry has increased notably over the past decade, with large store formats gaining an increased market share. However, small-format stores have survived (e.g., convenience stores for top-up shopping) (Bell, 2003; Dobson, 2003). Groceries are low outlay, frequently purchased, and tangible products with varying differentiation potential (Peterson et al., 1997). Since many grocery items are perishable, selling groceries online represents a specific challenge. Where home delivery is offered, distribution costs are likely to be relatively high compared to purchase volume and profit margin.

Internet Impact on the Grocery Retail Industry

Internet-induced change in the grocery retail industry has remained up to now far below expectations. In 1998, it was predicted that the market for online grocery sales would increase to U.S. $34 billion by 2003 (Palmer et al., 2000). More current estimates predict total sales of up to U.S. $3.5 billion in 2003, and none of the chains expects the share of online sales to exceed 5% of total sales over the long haul (Lee, 2003). The high number of failures of Internet pure players in grocery retailing emphasizes the importance of multi-channel strategies, which are now the dominant approach in online grocery retailing.

Four Cases from the Grocery Retail Industry

Four cases from the grocery retail industry have been chosen to illustrate the four multi-channel strategies introduced in the previous section. Case selection has

Table 1a. Case overview

	General Business	Online Activities
ALDI	• World's 18th largest retailer by sales. • More than 5,000 stores in 13 different countries on three continents, core market is Germany. • Consistently and vigorously pursues a discount strategy and focuses on limited assortment of private label products at low prices and good quality (Ewing et al., 2004). • Core range of standard groceries and non-food household items, additional "Special Purchases" line that changes once or twice a week.	• The Internet is mainly used as promotion channel aiming to increase allocation efficiency of the traditional channel. Placing orders online is not possible. • The site is scarce in content and reflects only a small subset of the offline assortment. Main focus of the site is information on the Special Purchases. • It has no significant influence on processes or transactions in the physical outlets. The only online marketing tool is a weekly e-mail newsletter. • http://www.aldi.com, http://www.aldi.de|fr|nl|...
AHOLD/ PEAPOD	• World's 3rd largest retailer by sales. • In 2002 sales of almost USD 60,000 Mio (Ahold, 2004; n.a., 2003). Operation of 5,606 stores through its subsidiaries on four continents. • Ahold USA heads six traditional retail companies with regional foci and Peapod, an Internet supermarket (Ahold, 2004).Peapod is an independent company with its own brand and Web site and provides online grocery shopping in five metropolitan areas in the US (Peapod, 2004).	• On Peapod's Web site, a full-fledged online supermarket is offered. Peapod cooperates with Ahold's Stop & Shop and Giant Food markets, whose Web sites do not offer online shopping, but provide a link to Peapod. • In the Chicago region in contrast, Peapod operates independently of Ahold's other chains and operates own infrastructure, incl. warehouses and distribution centers. • Peapod pursues a hybrid model using own as well as Ahold's chain store infrastructure. • http://www.peapod.com http://www.giantfood.com, http://www.stopandshop.com

Table 1b. Case overview

	General Business	Online Activities
LESHOP	• Switzerland's first online supermarket offering home delivery all over the country, founded 1997. • After a promising take-off, it was close to being shut down in December 2002 (LeShop.ch, 2004). However, LeShop's management was able to find a group of private investors and succeeded in continuing all activities (Bon appétit Group, 2003). • LeShop's turnover is growing continually and reached CHF 14.7 Mio in 2003. It claims to have a client base of 16,000 regular customers at present (LeShop.ch, 2004).	• In 2003 LeShop announced a strategic alliance with Migros, Swiss market leader in grocery retailing. • Both Migros' and LeShop's Web sites now give access to the same online supermarket. Migros' infrastructure is combined with LeShop's proprietary distribution centre. The new shop went online on January 19, 2004, offering a large range of products (6,000 items). • Migros and LeShop jointly promote the new online shop. Le Shop benefits from Migros' strong brand in the traditional retail market (LeShop.ch, 2004). • http://www.migros-shop.ch, http://www.leshop.ch
TESCO	• Tesco is the UK's leading grocery chain and was ranked 8th by sales of the world largest retailers in 2002 (n.a., 2003). • In early 2003, Tesco operated 2,291 stores in 10 different countries (Maddali, 2003). • Tesco struggled in the early 1990s, but success-fully relied on sales of non-food items and inter-national expansion. The strategy was redefined to a customer centric approach with the core purpose of creating value for their customers to earn their lifetime loyalty (Tesco, 2003). • Tesco's loyalty scheme "Clubcard" has become an inter-industry benchmark for customer profiling and CRM (McKelvey, 2003).	• Tesco launched its online channel Tesco Direct in 1996. Until 2000 the offer was limited to groceries and only run on a regional basis. Tesco Direct now offers a wide variety of products. • Tesco is considered the world's largest and most profitable Internet grocer (Seybold, 2001). Each store's pricing and inventory system is linked to tesco.com. Customers can select goods with prices they are used to. Orders are composed in local stores and delivered in customer defined two-hour delivery slots (Seybold, 2001). • Channel integration at Tesco is complemented by the Clubcard scheme. The cross-channel customer data is used for personalized marketing activities. • http://www.tesco.com

been instrumental (Stake, 1994), since the aim of the sampling process is not to prove generalizability, but to illustrate that the identified strategies exist empirically almost in a pure form. Furthermore, we want to give evidence of the practical relevance and the existence of the conceptually defined strategies even within one and the same industry. An overview of the cases can be found in Tables 1a and 1b.

Classification and Analysis of the Cases

Using the framework of multi-channel strategies, the cases introduced in the previous section will be analyzed. Since they relatively closely resemble the ideal types of strategy defined within the framework, they contribute to an explanation of the strategy types. Thereafter, we will conduct a more detailed case analysis based on indices that reflect the different approaches to the implementation of the multi-channel strategies in the online channel.

Classification of the Cases from the Grocery Retail Industry

The four cases closely resemble one of the polar types of strategies included in the framework. They further explain and give evidence of the strategies' relevance.

ALDI: Offline-Dominated Strategy

ALDI pursues an offline-dominated strategy. ALDI is positioned as cost leader and follows a low-complexity, no-frills approach with small assortments. The distribution system has been optimized for store delivery. ALDI's customers tend to be price-sensitive customers, and it is unlikely that they would be willing to pay a premium for multi-channel services such as in-store pickup or home delivery. Due to ALDI's low per-item-margin, providing these services without a price premium probably would be too costly.

AHOLD/PEAPOD: Isolation Strategy

With its supermarket chains and PEAPOD as an online grocer, AHOLD pursues an isolation strategy in the U.S. market. Two distinct supermarket chains are used as a physical basis for PEAPOD's online business. Thus, AHOLD extends its online grocery business without having to reposition its supermarket chains as multi-channel retailers. Instead, AHOLD benefits from and further strengthens PEAPOD's reputation as an online retailer. Involving the supermarket chains in the fulfillment of PEAPOD's online business provides PEAPOD with low cost access to infrastructure. Channel conflict potential is lowered, since the supermarket chains are involved in the online activities rather than being completely bypassed. While an often proclaimed advantage of Internet-based commerce is the geographical extension of existing offline-dominated businesses, AHOLD uses its store infrastructure to extend the geographical reach of its online business.

LESHOP: Online-Dominated Strategy

By teaming up with Migros, Internet pure player LESHOP increased its offline presence through joint promotion activities and by getting access to MIGROS' range of private label products. Furthermore, LESHOP can benefit from MIGROS' high brand awareness. The promotion activities and the reputation effects from collaboration with a renowned retailer aim to guide more customers to LESHOP's Web site. By sharing the online store infrastructure and an offline distribution infrastructure, costs are reduced.

TESCO: Integration Strategy

TESCO is positioned as a service-oriented retailer, which is offering low as well as upscale products. It pursues an integration strategy. TESCO's Web site offers online shopping, home delivery, and extensive information services. Most functions require a customer login based on the Clubcard number. Supermarkets are used to inform customers (i.e., through leaflets and messages printed on the customers' receipts) about the possibility and the convenience of online shopping and home delivery. However, TESCO does not enable customer channel switching at any point during a transaction. Features such as in-store pickup of shopping baskets ordered online are not offered.

Operationalization of Multi-Channel Strategies in the Web Channel

The implementation of click and mortar strategies encompasses virtual as well as physical channels. Online and offline channels can thereby be configured differently with regard to their support of transactions in the other channel, as illustrated by the framework and the four cases. Business strategies are typically a conglomerate of goals, plans, and policies. Due to the competitive nature of most strategies, companies face a dilemma of how much they want to reveal to their customers in order to properly set and manage expectations, and how little they want to reveal to their competitors in order to maintain their competitive edge and a moment of surprise. Hence, they are difficult to observe and comprehend from an outside perspective. In order to properly assess and describe a company's strategic rationale, one would have to include internal information on the actual process of strategy definition. Yet, we believe that particularly in the area of marketing channel definition, the external view can deliver valuable insights in the strategic design of a company's marketing system, because it is directed primarily toward customers. Here, the strategy implementation is most likely observable as it is reflected in the characteristics of the physical and virtual customer interface. Thus, we focus on the features and characteristics of a company's online channel and its publicly accessible information when we evaluate the multi-channel strategies of the four cases. In this stage of our research, the classification of strategies (Figure 1) as a result of phenomenological and literature-based work needs to be operationalized. We define 27 items that describe the underlying constructs and can be used to assess companies' Web sites. The items are thematically grouped according to (online) transaction phases (Selz & Schubert, 1997). For all items, a five point Likert scale is used. For the values 1, 3, and 5, a short description is given for clarification and reliability.[2]

In order to evaluate and analyze the implementation of multi-channel strategies, we have developed a set of indices. It describes the basic dimensions used in our framework (Figure 1): (1) the degree of online transaction support by the Web site, (2) the degree of offline support by the Web site, and (3) the derived level

Table 2. Description of dimensions and index compositions

Index name	Description
offline support	Features and information presented on a Web site which contribute to the support of transactions in physical outlets of the company Items: *OUTLETINFO, OUTLETCOMM, PRODINFO-OFFLINE, OFFLIAVAIL, OFFLIASSI, SHOPORDER, SHOPRETURN*
online support	Features and information presented on a Web site which contribute to the support of transactions on the site itself as virtual outlet of the company Items: *PRODINFO, ONLIAVAIL, ONLIASSI, HOMEORDER, DIRECTRETURN*
integration	Features and information presented on a Web site which lead to the perception of an integrated multi channel system in a sense that customers are enabled or encouraged to freely switch channels within or between transactions. This index combines the previous two indices with additional items. Items: *OUTLETINFO, OUTLETCOMM, BRANDS, PRODINFO, PRODINFO-OFFLINE, ONLIAVAIL, OFFLIAVAIL, ONLIASSI, OFFLIASSI, HOMEORDER, SHOPORDER, DIRECTRETURN, SHOPRETURN*
customer retention	Features and information presented on a Web site which indicate a focus on (online) customer retention in the marketing strategy of the company. Items: *ACCOUNT, REGNEED, ITEMALERT, OFFERALERT, CONTENT, FLEXPRICE, ORDERSTATUS, LOYALTY*
communication	Features and information presented on a Web site which denote a strong and open information policy held by the company. Items: *GENINFO, GENCOMM, OUTLETINFO, OUTLETCOMM, BRANDS, ONLISCOPE, ROLECOMM, REGEXPL*

of online/offline channel integration. These are theoretical constructs allowing us to identify a relative positioning of the physical and virtual channel as the result of strategy implementation. The indices are additive and not weighted, because, at this stage of research, we see no indication for assigning different weights to certain items. Table 2 gives a description of the constructs and indices. The index composition is also disclosed by naming the item codes as used in the survey instrument.[2] The allocation of items to indices follows a normative expert judgment at this stage of research.

The two additional indices—customer retention and communication—defined in Table 2 are given for illustration purposes only. They show the additional potential of our evaluation approach and were composed to further describe and assess the functional scope and implementation quality of multi-channel strategies. Based on the indices we will formulate exemplary research propositions regarding interdependencies between the retention and communication indices and the type of the multi-channel strategy.

For customer retention, we propose that a high level of online customer retention measures will be found wherever the online support index has a high value. For communication, we propose that the more a company supports transactions in any channel through the Web site, the more online communication measures can be expected. With increasing transaction support, an online customer interface becomes more complex, and the lack of physical contact and communication options has to be compensated for. In integration strategies, we expect the highest level of communication and information measures, since intensive information on both online and offline channels should be provided.

Despite methodological limitations, which are discussed later, we found that this approach leads to valuable insights. One reason is that the Internet can be used to easily provide information and communicate with customers. Experience shows that companies tend to more or less openly communicate the roles of physical and virtual channels in their multi-channel strategy on their Web sites. Thus, it is possible to identify and illustrate fundamental differences between the different strategies followed, even though the online channel is the sole source of information.

Findings and Discussion

Figure 2 shows an illustration of the index values calculated for the four multi-channel cases. Vertical positioning in the diagram depicts the index value measuring the degree of online transaction support by the Web site. Horizontal positioning depicts the index value measuring the degree of offline transaction support by the Web site.

Figure 2. Evaluation results

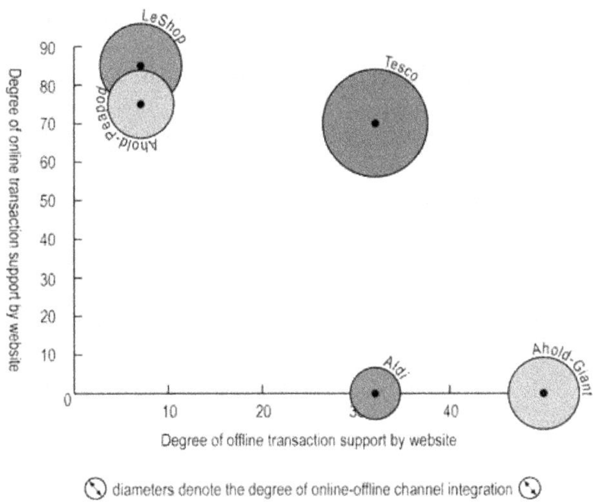

The index values on the vertical and horizontal axes as well as the diameters are percentages, which are calculated by relating the additive unweighted index scores to the maximum score that could be achieved for each index.[3]

The diameters denote the degree of offline/online integration. This means that in cases where the Web site extensively supports both offline and online transactions and thus resembles an integration strategy, the diameter will be larger. In cases where transactions are only supported to a low extent in either channel, the diameter is smaller.

ALDI's offline-dominated strategy is clearly reflected by the Web site's positioning in the diagram. While offline transactions are only partially supported, there is no support of online transactions at all, which again leads to a low integration index value reflected in the small diameter.

Compared to ALDI, LESHOP's online-dominated strategy is based on an almost inverse channel configuration: the Web presence is LESHOP's only customer interface, which is also marketed to MIGROS' customers. Functionality of LESHOP's Web site for online transaction support is more extensive than ALDI's for offline transaction support, which explains the relatively high value for LESHOP's online channel transaction index.

The diagram also illustrates AHOLD's isolation strategy. This strategy is reflected in the approach of multiple online presences for retail chains such as Stop and

Shop and GIANT, and for the online business PEAPOD. While GIANT's Web site as a representative of the offline-dominated part of the business provides more information and customer support than ALDI's Web site, online transaction support is not available here. Instead, online transactions have been isolated from offline activities and were implemented under the PEAPOD brand name on a different Web site. Consequently, following the isolation approach, it only supports online transactions. It is used to capture online sales potential from regions that are covered by different AHOLD retail chains as well as regions only served through the online channel.

TESCO uses its Web site to support online and offline activities. On the one hand, the Web site is used to offer customers a wide range of groceries, additional products, and services. On the other hand, the Web also is used to support offline transactions (e. g., providing information on retail outlets) and an integrated loyalty scheme for online and offline channels. TESCO's Web site shows a balanced support of online and offline activities, facilitating customers' shopping experiences in either channel, which leads to a high level of channel integration.

Implications for Multi-Channel Strategies

The indices used to analyze and illustrate the multi-channel strategies in the previous section focused mainly on transactional features that enable or support customer online or offline transactions. However, the use of Web sites by retailers is not limited to features directly supporting these but can comprise further elements supporting communication with and information of customers. A Web site also can contain different elements aiming to increase customer retention. We propose that the extent to which a Web site is used for communication or for customer retention purposes varies, depending on the chosen multi-channel strategy. In the following, we will discuss these propositions and discuss the extent to which our four cases provide first evidence for them.

Customer Retention Measures Implemented on the Web Site

The online channel and interface design can be used to lock in customers to a certain extent (Riemer & Totz, 2003) and, thus, to increase customer retention. By offering customer accounts with an extensive functionality that allows customers to define personal profiles, customers' decisions to switch to other retailers potentially can be influenced, since the personalization of information and account services increases switching costs. Typical examples for retention

Figure 3. Customer retention index values

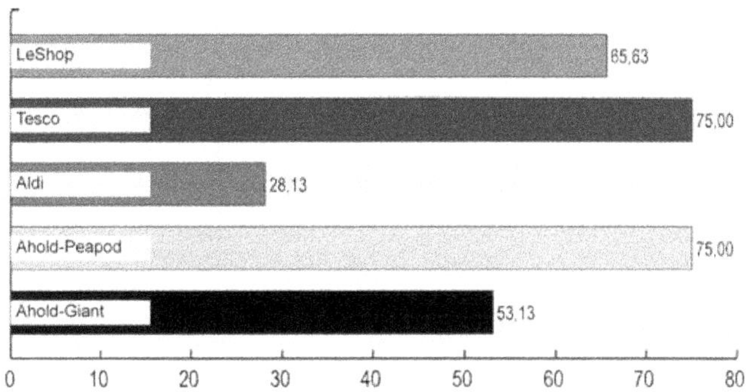

measures are the integration of a (cross channel) loyalty program, order status information, or additional content that is more contextual than directly product-related and is updated regularly in order to attract customers to the Web site on a regular basis. We propose that the extent to which these features can be found on a Web site will vary, depending on the multi-channel strategy chosen by a company. Wherever the online support index has a high value, we expect a high level of online customer retention measures. We used the index to find out whether our four cases provide evidence for this proposition. The results are illustrated in Figure 3.

LeShop and Peapod rank high, as expected. In both cases, the online channel is the only distribution channel, so the intense use of online customer retention measures makes sense. Tesco takes a prominent position, as well. This can be explained by the strong emphasis that Tesco puts on customer relationship management, which is also consistently practiced and executed on the Web. Thus, the three strategies in which the online channel plays a major role for transaction purposes also implement extensive customer retention features.

Aldi has hardly implemented any features aimed at supporting customer retention. This is consistent with the general role of the online channel, which is mainly used as a means to attract customers to retail outlets. In the case of Giant however, where an offline-dominated strategy is followed, as well, a more extensive use of customer retention features has been implemented. This difference cannot be explained simply through the different types of multi-channel strategies. Rather, it seems that in addition to the chosen multi-channel strategy, the general marketing strategy of the two retail chains influences the extent to which customer retention measures have been implemented on the Web site. Aldi follows a no-frills, cost-leader-oriented strategy, selling to anonymous customers. GIANT and AHOLD, in contrast, follow a more service-oriented approach, using marketing instruments such as a loyalty scheme in order to recognize customers on an individual level, even in the offline domain.

Scope of Online Communication and Information

The Internet presents the opportunity to provide information on various issues in a cost-efficient way and to support communication with customers. Retailers' Web sites, for instance, can be used to support communication both as a central point of contact and via retail outlets. Customers can be given the opportunity to give feedback. Furthermore, information on the company in general, on individual retail outlets, on how to use the Web site, and on a Web site's role within a physical-virtual multi-channel system can be provided. We propose that the more a company supports transactions in any channel through the Web site, the more online communication measures can be expected. With increasing transaction support, the online customer interface becomes more complex, and the lack of personal contact and communication options has to be compensated for. In integration strategies, we expect the highest level of communication and information measures, since intensive information on both online and offline channels should be provided. Again, we used the index to find out whether the four cases analyzed provide evidence for this proposition. The results are illustrated in Figure 4.

The four cases provide first evidence for our research propositions. TESCO offers the most extensive information on its Web site. This can be explained by the fact that information on activities in both online and offline and offline channels is provided, and customers are given the opportunity to interact with the company.

Figure 4. Communication index values

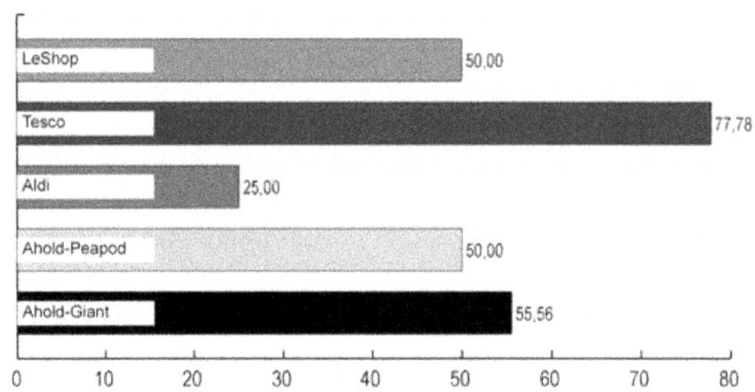

Three of the Web sites that only contain information on one channel offer about the same amount of information: LeShop's Web site focuses on the online channel only. Due to Ahold's isolation strategy, Peapod's Web site does not contain information on Ahold's retail store chain outlets, while Giant's Web site focuses on its retail outlet business. The difference between Giant and Aldi can be explained by looking at the different degrees to which offline transactions are supported, which is also reflected in the different positioning in Figure 2.

Limitations of Research

Our study of multi-channel strategies and their implications has been limited to the Web channel. While it has given evidence that it is possible to study multi-channel strategies by focusing on the Web site, some limitations also have become obvious. The four cases were useful to explain and gain insights into the four distinct types of multi-channel strategies included in our framework. Yet, we expect some difficulties when classifying multi-channel strategies only by focusing on Web site analysis. In cases where a click and mortar retailer operates a Web site supporting complete online purchasing transactions but has no separate Web site for outlet-based activities, as in the case of Ahold, it might be hard to determine whether the retailer follows an online-dominated or an isolation strategy. Further information, probably on internal issues such as the integration of operational processes and coordination between physical and virtual channels, might be needed. Furthermore, our research has so far been of an exploratory nature only. Our empirical evidence has been selected based on an instrumental sampling approach, in which cases were selected to explain and illustrate certain issues. The applicability of the framework and the approach we took to operationalize the strategies in other industries and to other cases still have to be tested.

The concentration on Web-based analysis in studying multi-channel strategies leads to further limitations. We argue that because companies increasingly use the Web as a core channel for customer communication, this is a legitimate way of studying companies' strategies. However, this approach limits our ability to distinguish between strategic decisions and their implementation. We take the implementation of strategies, as they are observable on the Web for face value.

Conclusion and Outlook

In contrast to a section of the literature that normally suggests that companies should pursue a multi-channel integration strategy, we have identified four basic strategies that are consistent within a given set of strategic assumptions. We have identified instances of each of the four strategies and have explored the differences regarding the implementation of customer retention and online communication as core elements of multi-channel strategies. While the analysis of customer retention measures implemented online did not yield results that were consistent with our expectations, the results regarding the communication and information features on the retailer's Web site did. For management purposes, the framework of multi-channel strategies and the defined variables contained in the developed indices can be used in a heuristic manner, since they give an overview of the potentialities regarding strategic positioning of multi-channel activities and of various implementation options. We also have demonstrated in an exemplary way that the framework can be used to derive research propositions and hypotheses, thus pointing out potential directions for future research.

Acknowledgments

This publication is based on work done within the research cluster Internet Economy at the University of Münster. The authors wish to thank the German Federal Ministry of Education and Research (BMBF) for financial support (grant number 08 NM 065 A).

References

100 largest global retailers. (2003). Retrieved January 22, 2004, from *http://www.chainstoreage.com/industry_data/pdfs/global100/2003_globaltop100.pdf*

Ahold. (2004). Company profile. Retrieved January 28, 2004, from *http://www.ahold.com/aboutahold/companyprofile.asp*

Anderson, E., Day, G.S., & Rangan, V.K. (1997). Strategic channel design. *Sloan Management Review, 38*(4), 59-69.

Bell, R. (2003). Competition issues in European grocery retailing. *European Retail Digest, 39*, 27-37.

Bon Appétit Group. (2003). LeShop.ch, rescued by its own clients! Retrieved January 29, 2004, from *http://www.bon-appetit.ch/?wm =sp%28%2d1%2c6 %2c172c86%29c%28100%29*

Coughlan, A.T., Anderson, E., Stern, L.W., & El-Ansary, A.I. (2001). *Marketing channels*. Upper Saddle River, NJ: Prentice Hall.

Deighton, J. (1997). Commentary on "exploring the implications of the Internet on consumer marketing." *Journal of the Academy of Marketing Science, 25*(4), 347-351.

Dobson, P.W. (2003). Competition and collaboration in European grocery retailing. *European Retail Digest, 39*, 13-21.

Doolin, B., & McQueen, B. (2003). Internet strategies for established retailers: Five cases from New Zealand. *Proceedings of the 16th BLED eCommerce Conference*.

Earl, M.J. (2000). Evolving the e-business. *Business Strategy Review, 11*(2), 33-38.

Ewing, J., et al. (2004). The next Wal-Mart? *Business Week*.

Gadde, L.E., & Ford, D. (2000). The role of distribution channels in value creation. *Proceedings of the International Marketing Educators Conference*, Buenos Aires.

Gallaugher, J.M. (2002). E-commerce and the UNDULATING distribution channel. *Communications of the ACM, 45*(7), 89-95.

Gulati, R., & Garino, J. (2000). Get the right mix of bricks & clicks. *Harvard Business Review, 78*(3), 107-114.

Lee, L. (2003). Online grocers: Finally delivering the lettuce. *Business Week*, 67.

LeShop.ch. (2004). Corporate information. Retrieved January 29, 2004, from *http://info.leshop.ch/uk/Welcome.html*

Maddali, P. (2003). Tesco in 2003. *Global CEO, 6*, 65-79.

Malone, T.W., & Crowston, K. (1994). The interdisciplinary study of coordination. *ACM Computing Surveys, 26*(1), 87-119.

McKelvey, C. (2003). How Tesco still rules the roost. *Precision Marketing, 15*(36), 15.

OC&C. (2002). Multichannel retailing: Der deutsche Einzelhandel steht noch am Anfang. Retrieved October 23, 2002, from *www.occ-strategy.de/resources/uploads/docs/multichannel.pdf*

Palmer, J., et al. (2000). Online grocery shopping around the world: Examples of key business models. *Communications of the AIS, 4*.

Peapod. (2004). Corporate fact sheet. Retrieved January 28, 2004, from *http://www.peapod.com/corpinfo/peapodFacts.pdf*

Peterson, R.A., Balasubramanian, S., & Bronnenberg, B.J. (1997). Exploring the implications of the Internet for consumer marketing. *Journal of the Academy of Marketing Science, 25*(4), 329-346.

Rheault, D., & Sheridan, S. (2002). Reconstruct your business around customers. *Journal of Business Strategy, 23*(2), 38-42.

Riemer, K., & Totz, C. (2003). The many faces of personalization—An integrative overview of mass customization and personalization. In M.M. Tseng, & F.T. Piller (Eds.), *The customer centric enterprise: Advances in mass customization and personalization* (pp. 35-50). New York: Springer.

Schoenbachler, D.D., & Gordon, G. L. (2002). Multi-channel shopping: Understanding what drives channel choice. *Journal of Consumer Marketing, 19*(1), 42-53.

Selz, D., & Schubert, P. (1997). Web assessment—A model for the evaluation and the assessment of successful electronic commerce applications. *International Journal of Electronic Markets, 7*(3), 46-48.

Seybold, P.B. (2001). Get inside the lives of your customers. *Harvard Business Review, 79*(5), 80-89.

Stake, R.E. (1994). Case studies. In N.K. Denzin, & Y.S. Lincoln (Eds.), *Handbook of qualitative research* (pp. 236-247). Thousand Oaks: Sage.

Steinfield, C., Bouwman, H., & Adelaar, T. (2002). The dynamics of click-and-mortar electronic commerce: Opportunities and management strategies. *International Journal of Electronic Commerce, 7*(1), 93-119.

Steinfield, C., et al. (2001). Pillars of virtual enterprise: Leveraging physical assets in the new economy. *info, 3*(3), 203-213.

Tesco. (2003). The Tesco strategy. Retrieved August 13, 2003, from *http://81.201.142.254/companyInfo/businessStrategy.asp?section=1*

Venkatesh, A. (1999). Virtual models of marketing and consumer behaviour. *Proceedings of the ESRC Virtual Society Workshop—E-Commerce and the Restructuring of Consumption,* London.

Webb, K.L. (2002). Managing channels of distribution in the age of electronic commerce. *Industrial Marketing Management, 31*(2), 95-102.

Wirtz, B.W. (2001). *Electronic business.* Wiesbaden: Gabler.

Endnotes

1 For a good synopsis of the evolution and structure of the grocery retailing industry, see Palmer et al. (2000).

2 Information on the survey instrument containing descriptions of the items can be requested from the authors.

3 E.g.,

$$\text{online support}_{TESCO} = \frac{\begin{array}{c} PRODINFO_{TESCO} + ONLIAVAIL_{TESCO} + ONLIASSI_{TESCO} + \\ HOMEORDER_{TESCO} + DIRECTRETURN_{TESCO} \end{array}}{5+5+5+5+5} = \frac{4+3+4+5+3}{5+5+5+5+5} = 0.7$$

5 is the maximum value that can be achieved for each individual item.

Chapter XIV

Describing the Critical Factors for Creating Successful Mobile Data Services

Anne Tseng, Helsinki School of Economics, Finland

Jukka Kallio, Helsinki School of Economics, Finland

Markku Tinnilä, Helsinki School of Economics, Finland

Abstract

Mobile operators play a central role in the development of the mobile data services market. They have primary access to the customer relationship, a key source of revenue, and are responsible for how revenue is distributed to other participants in the value chain. As a result, a successful operator-driven business model is essential to the survival of the mobile data industry. The purpose of this chapter is to describe the critical factors that have influenced the results of operators based on countries that have been at the forefront of mobile data services innovation. Then, by comparing the key characteristics of operator-driven business models in these four cases around the world, we will describe the critical factors used in designing successful mobile data services.

Introduction

Given the high penetration of mobile phones and the PC Internet, it has long been predicted that mobile data usage would increase substantially as a result of the intersection of the two channels. Unfortunately, however, mobile data usage has been slow to materialize. Despite the business potential, entrants and incumbents alike have been confounded by a host of unexpected challenges such as insufficient demand, competition from substitutes, and, most important, lack of profitable business models. In this chapter, we will attempt to address the last issue by conducting an exploratory study of the development of business models in Japan, Western Europe, South Korea, and China in order to describe the critical factors used in successful operator business models for mobile data services.

For the purposes of this analysis, we define *mobile data services* as any mobile non-voice service. This includes wireless data transfer technologies such as instant messaging and SMS as well as e-mail and the mobile Internet. We define *mobile Internet* as Internet access using mobile devices including but not limited to cell phones, personal digital assistants, and so forth. Moreover, since creating and capturing value is one of the most challenging issues in the mobile data services business, we shall use increased Average Revenue Per User (ARPU) as a key objective for mobile operators.[1]

Analysis By Country

Mobile operators play a central role in the development of the mobile data services market. They have primary access to the customer relationship, a key source of revenue, and are responsible for how revenue is distributed to other participants in the value chain. As a result, a successful operator-driven business model is essential to the survival of the mobile data industry. In the following four empirical cases, we will study the experiences of four operator groups that constitute some of the largest and most active mobile data service markets in the world; namely, the countries/regions of Japan, South Korea, China, and Europe.

We begin the discussion by comparing the success stories of selected operators in countries that have the highest mobile data services revenues. Mobile data revenues are the highest in Japan at 19% of total operator revenues, followed by Western Europe, where data revenues represent 11% of total revenues, most of which is SMS.[2] Korea comes in third with mobile data representing 9% of South Korean operators' revenues. China comes in fourth with 1% (see Figure 1).

Figure 1. Data as percentage of operator revenues

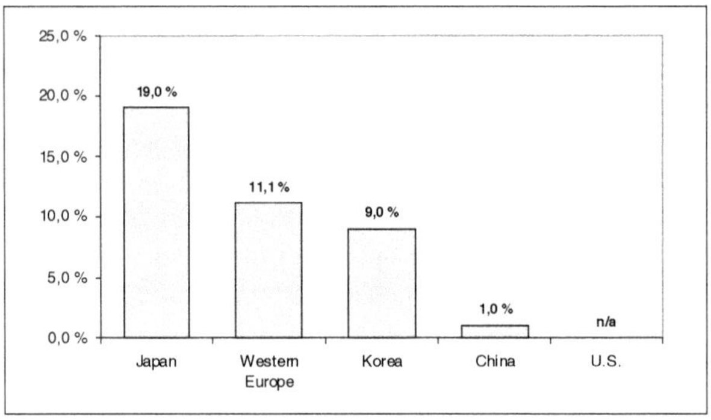

Source: Merrill Lynch Telecom Services Research, Wireless Matrix 3Q02, 10 Dec 2002 issue

Japan

With NTT DoCoMo's launch of i-mode in February 1999, Japan became the first country to successfully introduce a mobile data service and, with the exception of countries like South Korea, has been one of the few countries able to grow and sustain a significant subscriber base in mobile data services. As of March 2003, there were 61.8 million subscribers of mobile data services in Japan (Henten et al., 2003).

The operator made several departures from industry-accepted norms in its introduction of i-mode, which ultimately proved instrumental to its success. First, in contrast to the European operators who targeted mobile Internet products and service offerings toward the high-end business user, NTT DoCoMo wanted to create a product that would attract the average consumer. To do so, the design team insisted that the product be easy to use and affordable (Matsunaga, 2000). This affected the carrier's marketing approach, which did not focus on the technology, as was done with WAP in Europe, but rather on the user benefits of the i-mode service (Natsuno, 2000). Second, as early adopters, the operator targeted younger segments of the market to help initiate the positive network cycle necessary to spur mass consumer demand (Ratliff, 2002). In keeping with this segment, subscription prices were kept at impulse-buy levels in order to compete with magazines (Ratliff, 2002).

Most importantly, NTT DoCoMo viewed i-mode as a complete product. In order to implement its vision, DoCoMo played a key role as a coordinator of the value chain. By deciding every detail from the design of the handsets to what constituted official content on its mobile portal, the company ensured that its phones, applications, and Web sites were fully functional at launch (Matsunaga,

2000). Japanese manufacturers tailor-made i-mode cell phone models to DoCoMo's exact specifications, including screen size, weight, battery life, and functionality (Matsunaga, 2000). DoCoMo also chose to partner with content providers instead of purchasing or creating the information, deciding that rather than provide content, it would serve as a gateway for quality content (Natsuno, 2000). This emphasis would later contribute to its successful branding as a quality service offering.

The decision to focus on quality content also dictated the company's choice of technologies. Therefore, instead of using WAP, which was at the time being touted by the world's largest handset manufacturers (Nokia, Ericsson, and Motorola) as the global standard for mobile data services, DoCoMo chose cHTML, a subset of the HTML language used by the fixed Internet (Matsunaga, 2000). This made the transition from the fixed Internet to mobile data services easier for content providers (Natsuno, 2000). The company also decided to use packet-switched vs. circuit-switched technology, a move that also allowed users to access the Internet without charging them for the time spent online (Henten et al., 2003). Furthermore, DoCoMo did not charge content providers to participate and offered instead a revenue-sharing agreement in which it kept 9% of the content fees generated for its billing service, and content providers received 91% of the content fees (Ratliff, 2002). The end result was a positive network effect that led to a substantial increase in subscribers and wireless content. Interestingly, NTT DoCoMo's competitors, KDDI and J-Phone, have been able to successfully copy the i-mode business model, despite the use of different technological platforms and target segments (Bohlin et al., 2003).

South Korea

South Korean mobile operators have followed business models that are similar to Japan's i-mode model (i.e., coordinate the value chain through strong relationships with handset manufacturers and content providers with a generous revenue sharing agreement between operators and service providers) (Kelly et al., 2003). What makes their success more striking is that they have succeeded, in spite of using different technologies and platforms (i.e., CDMA and WAP). Moreover, operators have proven to be innovative in dealing with government regulations. In 2000, when the Korean government banned handset subsidies, Korean mobile operators responded by segmenting their voice offerings into multiple branded products (Strand, 2002). This meant that children, teenagers, single women, business men, and the elderly all had their own branded products and distinct portals as well as separate promotions and pricing plans. The novel approach allowed operators to successfully differentiate their services on brand and price, and increased demand for mobile voice and data services.

Europe

Despite China's market size potential, as a region, Western Europe still represents the largest market with 300 million cellular subscribers (Merrill Lynch, 3Q, 2002).[3] However, aside from SMS, Europe has been unsuccessful in introducing the mobile Internet to the mass consumer. Few users adopted WAP, and GPRS adoption rates have thus far been relatively low.[4]

Mobile operators made several critical mistakes in the introduction of WAP services. The service was mismarketed as fixed Internet on the mobile phone, suffered from poor service and inadequate technical support, lacked content, and was expensive (Baldi et al., 2002). Moreover, the service was modular, which meant that users had to custom download applications as they needed them, making it frustratingly difficult and time-consuming for customers to use. While WAP has improved significantly since its initial launch three years ago, it continues to suffer from the stigma of being over hyped. The launch of GPRS networks should improve users' reactions to the new service as the network is packet-switched, which is more conducive to transmitting data services. However, the success of GPRS is far from assured, and Europe's choice of 3G, W-CDMA, remains unproven.

The relative success of the fixed Internet in Europe also has affected operators' approach to the mobile Internet. Fear of becoming a bit-pipe plus the pressure to recoup the high prices paid for 3G spectrum have resulted in higher consumer prices and revenue-sharing models that have been less generous for content providers in Europe than in Asian countries (Quigley, 2001). Another consequence has been the decision by some operators to develop content in-house, a practice that has proven costly and time-consuming (Maitland et al., 2003). Gradually, however, operators are learning to partner, which means allowing content and service providers to co-exist with them in the mobile data services value chain, offering more generous revenue sharing agreements and focusing more on creating a positive user experience than on selling faster and more advanced technologies. Operators like Vodafone have begun to successfully copy elements of the i-mode business model, such as partnering with content providers, sharing revenues, and working more closely with handset manufacturers (Salz, 2003). After being shut out of 2G, Japanese handset manufacturers are seeing a resurgence of business from European operators, and even Nokia, the largest handset manufacturer, has started to conform its handsets to operators' demands (Longino, 2003). Whether or not Europe will be able to grow its position in the mobile data services market will depend on its ability to develop sustainable, profitable business models. Vodafone has been among the leaders, releasing its own version of I-mode—Vodafone Live!—and has successfully attracted a million subscribers after only five months in service. Meanwhile,

Virgin Mobile's continued expansion into the U.S. and other countries provides a useful lesson in branding and targeting.

China

As a latecomer into the mobile data services market, China was able to benefit from the experiences of both Europe and Japan. Like Europe, SMS is extremely popular in China (Albright, 2001; China Unicom, 2003, Merrill Lynch, 2002). In the second half of 2002, some 23.7 billion messages were transmitted over the country's mobile networks (China Unicom, 2003; Merrill Lynch Research, 2002). Like Europe, China Mobile, the country's largest operator, was unable to attract subscribers to its WAP services; only 8,000 users had signed up within five months of launch (Yan, 2001). Then, in November 2000, the operator introduced Monternet, a mobile portal initiative, patterned after NTT DoCoMo's i-mode business model. Moreover, China Mobile made substantial investments to facilitate its use, developing a common platform to facilitate mobile data roaming capabilities and a standardized billing system, requiring an upgrade of mobile operations in every province and metropolitan area to support the new billing requirements (Sigurdson, 2002). Mobile service providers collect the full retail price for their services and then compensate the operator for using their network and billing services (Zhang, 2001). Following the Japanese and Korean model, the operator also introduced a generous revenue share agreement of 91% to the content provider and 9% for the operator, with an additional 6% collected by the operator in the case of credit issues (Yan, 2001). The subsequent success of the mobile portal has since helped the three largest independent portals to become profitable for the first time (Kurtenbach, 2003).

Describing Critical Factors for Introduction of Successful Mobile Data Services

As the experiences of selected mobile operators show, successful business models are essential to the development of the mobile data services industry. Services need to be developed closely with technologies in order to create a positive user experience that will attract and retain customers. In order to create a positive user experience, time, effort, and cost must be invested to shepherd products from development to implementation. This requires the joint efforts of manufacturers, network operators, content and services providers, and systems integrators. Based on our comparisons of successful operator business models, we will describe these critical success factors for mobile operators.

To increase ARPU, we begin by focusing on the universal levers that an operator can influence and use to its advantage. As the experience of NTT DoCoMo demonstrates, these levers include content/handset/interface design, network quality, pricing, billing, marketing, and customer support (see Figure 2). By changing these levers, the operator can affect the user experience so that he or she will increase the amount of minutes used and data downloaded, which, in turn, will lead to increased ARPU.

To encourage increased usage, users must have a positive experience. A Nielsen Norman report on WAP usability, published in December 2000, concluded that usability drives mobile data service adoption. The average consumer may be unforgiving, if services are found difficult to use. Operators have to focus on creating a positive user experience through continual innovation and compelling content and services. In particular, subscribers have been attracted to services that are easy to use, reliable, fast, entertaining/informative, and reasonably priced. These elements can be achieved best through the cooperation of the handset manufacturer, operator, third party systems integrators, and content and application providers. This implies that if there is any discord between the three

Figure 2. Achieving operator's objective

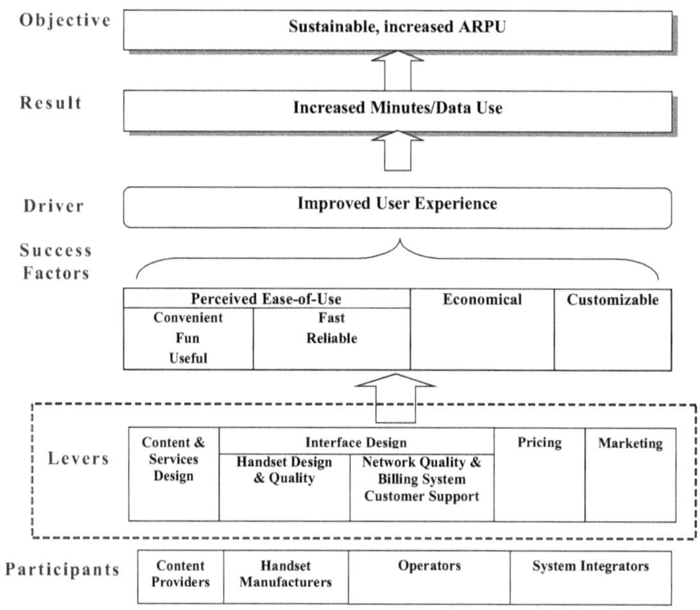

parties, then demand and the market for new content, and, hence, mobile data services will not develop. Thus, as NTT DoCoMo's experience suggests, the operator should focus on coordinating and controlling the quality of the final product produced as a result of the interaction between the different participants in the value chain. By doing so, the other components, such as technology, pricing, billing, and marketing, will fall into place (see Figure 2). Next, we describe several factors that may influence the successful deployment of mobile data services.

Content and Services Design

Operators who have been able to launch successful mobile data service offerings, like Vodafone Live! and i-mode, have learned that attracting quality content is imperative to attracting customers and increasing usage. NTT DoCoMo held its content providers to strict quality standards and acted as a certifier for its official sites. However, DoCoMo also tried to make the experience easier for the content provider. Its choice of cHTML as a platform instead of WAP had more to do with making things easier on the content provider than on how advanced the technology was.

The resulting content selection and quality has been astounding. There are 1,800 official sites accessible through i-mode, and services include photo exchanges, GPS location services, video-conferencing, and music/video/application downloads. The situation is similar in Korea, where even Korea's smallest carrier has some 300 providers and 5,000 services. Services include photo exchanges, GPS location services, video-conferencing, and music/video/application downloads.

Revenue Sharing

To attract a sufficient number of customers to their mobile data services business, successful mobile operators also need to work with a critical mass of content/service providers who can design innovative services, which, in turn, will attract customers. A revenue-sharing model that allows content providers to thrive and promotes healthy competition so that only the best services and content are delivered to the customer is crucial.

European mobile operators have been lambasted by content providers for exerting their market power to demand excess revenues from content providers in exchange for a good location on their mobile portals. Operators have been reluctant to share much of their own revenues because of their desire to improve ARPU. Worried they would become little more than a bit-pipe through which

Figure 3. Partners in the virtuous cycle

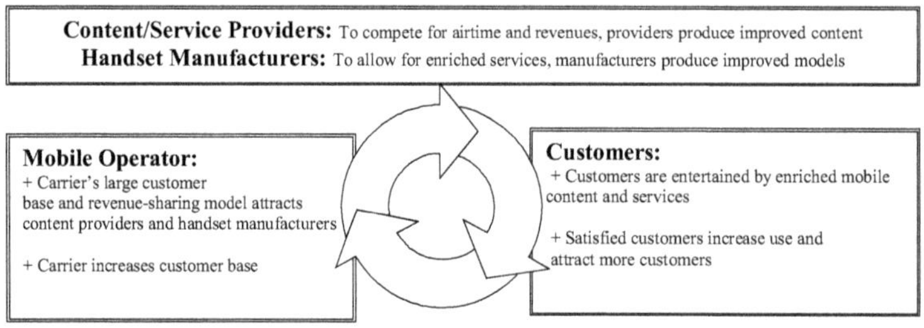

providers send their data, mobile operators attempted to control wireless Internet access by partnering with or developing their own services. This may be short-sighted, as the real revenue driver is in communication services (i.e., voice, e-mail, and data traffic) as opposed to content, so the operators should be creating an unattractive market in the long term for the sake of a relatively small short-term return. At the same time, however, operators may be more willing to share revenues, once wireless content services start to take off and content creators are perceived as value-adding.

Unlike their Western counterparts, Japanese operators do not want to be content providers and have not used their positions to demand an exess share of the revenue from data services. This has allowed all the parties involved to benefit from the revenue generated and aligns their strategies to a common goal of revenue expansion. As a result, content providers have been encouraged to invest in value-added services. DoCoMo calls this three-way relationship the virtuous cycle, with all four parties—subscriber, content provider, handset manufacturer, and operator driving the market (see Figure 3).

Marketing

With voice, operators have typically relied on two methods to segment the market: pricing and technology. However, with wireless data services, the rules appear to have changed. The operator needs to educate the consumer about the user benefits of new technologies (e.g., GPRS). The more successful operators have paid significant attention to marketing in order to educate the public about new services and to reach out to specific segments of the population who would be especially attracted to the service through targeted marketing approaches.

Teens are commonly and effectively targeted as a segment of the population most likely to be interested in mobile data services. This group includes teens that are supported by their parents as well as young adults in their 20s and 30s with high disposable incomes. They played a significant role in the eventual mass adoption of mobile data services in Japan and were credited with discovering and initiating the widespread adoption of SMS in Europe. Not surprisingly, Vodafone has chosen to target this segment with its new Vodafone Live service offering. In mobile voice services, Korean and Japanese operators also have used multiple brands to personalize users' experiences, thereby reducing the likelihood of churn and increasing revenue. Virgin Mobile has built its MVNO business on this segment, as well.

In South Korea, operators went even further, splitting their standard voice services into sub-brands, specifically targeted at different age and gender segments of the mobile users. These new sub-brands have their own portals, marketing, content and services offerings, and pricing, and have been instrumental in getting the mobile users to discard their 2G mobile phones and buy the new non-subsidised 2.5G mobile phones. The sub-brands have proven to be such a hit that only 24% of Korea's largest mobile operator subscribers remain on a standard price plan.

The way mobile data services are marketed also can impact perception and, therefore, its uptake. Operators have differed in the marketing messages they have used to reach out to potential users. Japan, Korea, and China have focused primarily on the entertainment aspect of mobile data services, stressing user benefits, whereas European operators have tended to focus on the technology, lower prices, prepaid plans, and switching operators in order to save money. U.S. operators have followed a similar path, highlighting the technical aspects of high-speed and always-on connectivity in addition to picture e-mail.

For example, Vodafone Live, in a departure from the customary marketing habits of European operators, is using trendy advertisements to sell its new service as fun and entertaining in attempt to make users aware of the fact that there is more one can do with his or her cell phone than text messaging or calling. This approach may have contributed to its exceptional customer uptake within the first five months of launch.

Equally important is making sure products are not hyped and that they perform as they are billed. For example, WAP service was billed as the fixed Internet over the phone, and users were led to believe that the mobile Internet would be as exciting as the fixed-line Internet and were disappointed when reality did not match expectations.

Pricing

While operators in the U.S. and Europe struggle to find a pricing structure that works for mobile data, Japanese and Korean operators provide some possible solutions. Japanese operators use a hybrid of subscription and per-data usage fees. I-mode and EZWeb uses low-cost monthly subscription fees in addition to a set data traffic fee per packet of data sent/received. J-Phone has no subscription but charges per data sent/received, a practice that Vodafone Live now is trying to copy.

Like J-Phone, Korean operators do not charge subscription prices; however, they go one step further—they distinguish between different types of data traffic. Users pay a lower rate for multimedia downloads, which tend to be data intensive but relatively low in value, than for text-based downloads, which are less data intensive, but have relatively high user value.

The advantages of these pricing plans are that they are relatively simple for users to understand and allow the carrier to further segment the market by price. However, they may be difficult to implement and may require expensive conversions from existing billing systems that have been based on billing for voice minutes as opposed to data transactions.

Billing Systems

Operators are finding that flexible billing systems are essential for the development of sophisticated services and business models and for maintaining customers. Once users are willing to pay for content, the method of payment becomes critical in ensuring the customer goes through with the sale. At this point, the success of content or service depends on the billing system. While European operators have found some success with reverse billing, many still have a long way to go before GPRS billing systems accurately capture the needs of all its customers (e.g., prepaid customers' real-time needs). U.S. operators are facing similar difficulties with legacy billing platforms that were built to track voice usage and can't easily or accurately account for the number of data transactions. Again, Vodafone has made advances with its M-Pay Service, which allows users to pay for low-cost digital and hard items valued at less than £5.00 and a pay-as-you-go system that allows it to offer mobile data services to its prepaid customers. In the meantime, Asian operators continue to make strides in the billing front. In 2002, Korean SK Telecom agreed to team up with LG Telecom to cooperate on offering mobile settlement and payment services in a preliminary step to open m-commerce opportunities. NTT DoCoMo upgraded its billing system so costs for premium services could be added to a user's phone bill, as did China Mobile as part of the launch of its mobile Internet portal.

Handset Design

The importance of handset design in enhancing the user experience should not be underestimated. Both Japanese and Korean operators have worked closely with handset manufacturers to create handsets that optimize user experience. As a result, Japanese phones are laden with features, function according to operator and manufacturer's specifications, and fit with the usage patterns of their target customer—the Japanese consumer. The latest handset models in Korea are among the most advanced in the world and can access the Internet at a rate up to 2.4 megabits per second, four times as fast as GPRS phones. This creates a virtuous cycle in itself as handset manufacturers continually improve handset technologies so that more technically advanced services can be offered. This, in turn, attracts customers to buy new handsets, replacing their older models and increasing traffic, thereby creating a win-win situation for both the operator and the handset manufacturer. In contrast, Western operators have a more contentious relationship with their dominant manufacturers, which include Nokia, Motorola, and Siemens, and they have had less influence over handset design than their Asian counterparts, which may be influencing the less-than-seamless quality of services that have been provided in Europe. This however, may soon be changing with Vodafone Live's precedent-setting move.

Network Quality

While increased bandwidth alone will not necessarily increase adoption, poor network quality can have a negative effect on user experience. Initial users of WAP experienced dropped data sessions, poor data throughput, and session instability as a result of poor network quality. In fact, U.S. networks suffer from poor voice quality due to the practice of compressing signals as a result of limited spectrum. Although GSM networks still dominate, operators have made inroads using CDMA 1x technology as an emerging 2.5G network. With the advantages of a lower cost to upgrade networks, better capacity and speed, and affordable and attractive handsets, CDMA networks offer distinct advantages over other types of networks at this time. W-CDMA deployment unfortunately is still very limited, handsets are rare, and technology is expensive. (GPRS has scale, but suffers from a limited number of color handsets, slower speeds, and the risk of cannibalizing voice capacity.) Table 1 compares the major levers/drivers that determine the different business models of operators around the world.

Table 1. Comparison of business models

Lever	Japan	South Korea	Europe	China
Content/ Service	1,800 sites from I-mode (800-900 each from J-Phone and KDDI). Services include photo exchanges, GPS location services, video-conferencing, and music/video/ application downloads	Korea's smallest carrier has 300 providers and 5,000 services. Services include photo exchanges, GPS location services, video-conferencing, and music/video/ application downloads	Limited - Vodafone Live! claims it has been able to attract 250+providers. Services include photo exchange, location-based services, application and ringtone downloads, SMS	Monternet reportedly has 130+ content providers
Revenue Sharing	Yes. Japanese operators pass 88-91% of subscription fees to content providers	Yes.	For Premium SMS mostly with European operators getting a larger cut than in Japan	Yes. Chinese operators relinquish 85-91% of content fee to providers
Marketing	Subbranding by age, gender, and corporate type	Subbranding by age, gender, and corporate type	Vodafone Live! and Virgin Mobile targeting youths or young adults	N/a
Marketing Message	Fun and entertaining services tailored to customer segments	Fun and entertaining services tailored to customer segments	Technology, low price, prepaid plans, and switching operators to save $$$	Free Inter-connections, Limitless Possibilities tailored services to customer segments
Pricing	Subscription fees and per-packet data traffic fees	No subscription; only per-packet data traffic fees	Per kilobyte under GPRS and per message for SMS	Subscriptions and data traffic fees
Billing Systems	Billed directly to mobile phone bill	Billed directly to mobile phone bill	Premium SMS bills directly to phone bill. Different macro- and micro-payment systems co-exist	Charged directly to mobile phone bill (China Mobile)
Handset Design	Domestic suppliers offer 65,000 color screen, 12-line displays, polyphonic speakers, digital cameras, and Internet browsers	Domestic suppliers offer 65,000 color screen, 12-line displays, polyphonic speakers, digital cameras, and Internet browsers	Nokia, Motorola, and entries from Panasonic and Sharp increase the number of camera phones	Nokia, Motorola, Samsung, Domestics gaining ground
Network Quality	Good – GPRS & PDC in Japan	Excellent – CDMA & WAP	Excellent-GSM/GPRS/WAP	Good-GSM/ GPRS/CDMA

Conclusion

Our findings suggest that a successful transition from mobile voice services to mobile data services represents a significant shift in focus and approach for all industry participants and, in particular, for operators. Instead of focusing purely on increasing bandwidth and complex technologies, successful operators are increasingly focusing on improving the user experience through coordinated handset and service design. They also are focusing on creating effective billing systems, offering services at reasonable prices, and targeting marketing strategies. We have described several factors that can influence the strategies of mobile operators that wish to implement successful mobile data services.

The achievement of successful operators, especially in Japan and Korea, can be distilled to this key phrase: Applications drive traffic and traffic drives revenue. In order to manage the process, successful mobile operators coordinate the relationship between different members of the mobile services value chain. By coordinating, the successful operator ensures that every participant has a clearly defined role in the value chain and is compensated for it. This includes fostering an open platform in which research and development can be shared, ensuring that participants' interests are aligned with the customer's (in a model that is essentially customer-driven), thus allowing all participants to maximize their value in the value chain.

These findings represent opportunities for further research in other issues surrounding mobile operator-driven business models. One possibility is to develop a new theory into which factors are more difficult to transfer and why. Continued research in this area could address the difficulties of copying successful operator-driven business models.

References

Albright, P. (2001). China may join SMS craze [electronic version]. *Wireless Week*. Retrieved May 20, 2003, from *http://www.wirelessweek.com/index.asp?layout=article&articleid=CA8236*1

Baldi, S., & Thaung, H.P. (2002). The entertaining way to m-commerce: Japan's approach to the mobile Internet—A model for Europe? *Electronic Markets, 12*(1), 6-13.

Bohlin, E., Bjorkdahl, J., Lindmark, S., & Burgelman, J. (2003). Strategies for making mobile communications work for Europe: Implications from a comparative study. *Proceedings of the European Policy Research Conference (EuroCPR)*, Stockholm, Sweden.

China Unicom. (2003). Announcement in respect of statistics data for the month of April 2003. Retrieved May 19, 2003, from *http://www.chinaunicom.com.hk/main/eng*

Henten, A., Olesen, H., Saugstrup, D., & Tan, S. (2003). New mobile systems and services in Europe, Japan and South Korea. *Proceedings of the Stockholm Mobility Roundtable 2003,* Stockholm, Sweden.

Kelly, T., Gray V., & Minges, M. (2003). Broadband Korea: Internet case study. *ITU*. Retrieved May 20, 2003, from *http://www.itu.int/ITU-D/ict/cs/korea/material/CS_KOR.pdf*

Kurtenbach, E. (2003). China's Internet companies roar back to life—Thanks to mobile phones. *Wireless Business & Technology Magazine, 3*(2).

Longino, C. (2003). Live!—An agent of change: Will Vodafone Live! shake up the mobile world beyond carrier portals? *The Feature: It's All About the Internet.* Retrieved May 20, 2003, from *http://www.thefeature.com/*

Maitland, C.F., van de Kar, E.A.M., & Wehn de Montalvo, U. (2003). Network formation for provision of mobile information and entertainment services. *Proceedings of the 16th Bled Electronic Commerce Conference e-Transformation*, Bled, Slovenia.

Matsunaga, M. (2000). *The birth of i-mode: An analogue account of the mobile Internet.* Singapore: Chuang-Yi Publishing Pte Ltd.

Merrill Lynch Telecom Services Research. (2002). *Quarterly update on global wireless industry metrics.* Merrill Lynch Global Securities Research & Economics Group. Global Fundamental Equity Research Department.

Natsuno, T. (2000). *I-mode strategy.* London: John Wiley & Sons, Inc.

Poulbere, V. (2003). Mobile data services: A market of 15.4 billion EUR in 2002 in Western Europe thanks to the rapid growth of SMS services (2/3). *IDATE Newsletter No. 223.* Retrieved May 20, 2003 from *http://www.idate.fr*

Quigley, P. (2001). Warning to operators: Overhaul or say goodbye. *Wireless Week, 7*(38), 15. Retrieved September 17, 2001, from *http://www.wirelessweek.com/index.asp?layout=article& articleid= CA159353*

Ratliff, J. (2002). NTT DoCoMo and its i-mode success: Origins and implications. *California Management Review, 44*(3), 55.

Salz, P. (2003). The right stuff the feature: It's all about the Internet. Retrieved May 20, 2003, from *http://www.thefeature.com*

Sigurdson, J. (2002). China as number one — in mobiles. *International Institute for Asian Studies Newsletter, 29,* 1.

Strand, J. (2002). Sub-brands can help European mobile operators become profitable. *Wireless Watch Japan.* Retrieved May 20, 2003, from *http://technologyreports.net/wirelessreport/?articleID=817*

Yan, X. (2001). Mobile data communications in China. *Proceedings of the Sixth Asia-Pacific Regional Conference of the International Telecommunications Society*, Hong Kong.

Zhang, N. (2001). The wireless virtuous circle emerges Linktone. *Proceedings of the I and I Asia,* Hong Kong, China.

Endnotes

[1] Although we will not focus on cost, it should be taken into consideration along with the business models, as many of these aspects will require significant investments in not only time and effort but also money.

[2] Western Europe in this context includes the countries of Austria, Belgium, Denmark, Finland, France, Germany, Greece, Italy, Netherlands, Norway, Portugal, Spain, Sweden, Switzerland, and the UK.

[3] This is evidenced by the three largest mobile operators, which are China Mobile followed by Vodafone (UK-based) and Deutsche Telekom (Germany-based).

[4] Even SMS, which generated an estimated • 11.7 billion of European operators' data revenues in 2002, was due more to chance discovery than through skilled marketing on the part of the operators (Poulbere, 2003).

Chapter XV

Channel Choices and Revenue Logics of Software Companies Developing Mobile Games

Risto Rajala, Helsinki School of Economics, Finland

Matti Rossi, Helsinki School of Economics, Finland

Virpi Kristiina Tuunainen, Helsinki School of Economics, Finland

Janne Vihinen, Helsinki School of Economics, Finland

Abstract

In this chapter, we explore the revenue logics and related product distribution models of mobile game developer companies. Mobile gaming is facing a transformation in both technical infrastructures and business models as it grows at a very fast pace. The former change originates from the technological shift of the environment of use; for example, from specific game consoles toward mobile phone platforms. The latter change relates to the possibility of delivering and playing games online, which affects both the distribution partnerships and the revenue stream options of mobile game vendors. We present a set of possible business models for game developers and concentrate on the possible combinations of revenue logics and distribution models for different games.

Introduction

The worldwide number of digital phone (GSM and PCS) subscribers has increased from 140 million in 1996 to approximately 900 million at the end of 2002 (GSMdata, 2002). Concurrently, the number of PC users is reaching a saturation point at around 400 million. From 2002 to 2004, the difference between phones and computers has continued to diminish with the arrival of Java-enabled phones and with a larger number of phones that support Web browsing and e-mail applications.

As mobile phones are rapidly turning into software platforms capable of supporting gaming, many handset manufacturers, operators, and game developers see the opportunity for mobile games. However, the recent downturn of investments into the enhanced cellular networks makes it challenging for companies to develop and deploy new advanced games. Furthermore, many aspects of the new business models, including revenue logics and distribution models for these new entertainment services, are still unproven. The mobile game market is expected to grow from $124 million in 2001 to exceed $4 billion in 2006 (Ovum, 2002). Today, most of the mobile gaming activity is in Asia-Pacific, particularly in Japan and South Korea, where there are tens of millions of subscribers of mobile entertainment services. However, we can expect that Europe and the US will soon see growth in these areas, as well.

Mobile games can be played with mobile phones; PDAs (Personal Digital Assistants), such as Palm or iPaq; Web-enabled phones; or other handheld game devices. In Europe, the development of mobile services has been characterized largely by technology push (Nurmi et al., 2001), but the future success of mobile services will strongly be affected by the ability of businesses to offer, already at an early stage, the right products and services to consumers (Anckar & D'Oncau, 2002). Experiences with PC-based Internet and Japanese mobile iMode services emphasize the role of entertainment services as a significant factor in the growth of mobile network usage. Games and entertainment services are important application areas for information industry as a whole (Shapiro & Varian, 1999), and, as the third generation mobile phone networks proliferate, demand for these services will increase rapidly.

In this chapter, we look at the mobile game scene and introduce a framework for analyzing software business models within it. We then develop the model further for mobile games and use it to discuss the revenue logics of mobile game developers. In the last section of this chapter, we summarize and draw conclusions on the discussed aspects of mobile games.

Types of Mobile Games

Generally, the existing games for mobile handsets are either server-based or stand-alone games. Server-based mobile games can be divided further into WAP, SMS, and Java games. Java games also can be used as stand-alone games. All of these games can be either single- or multi-player games. Multi-platform games, in turn, are a subset of games that can be played in conjunction with online, PC, and console versions.

A report of Durlacher Research (2001) suggests that mobile games can be classified by their operating and distribution platform into three types: stand-alone, server-based, and streamed. These games can be either downloadable from a server or preinstalled by a vendor or distribution partner.

- **Stand-alone games** do not require a network connection in order to play the game. As they run on mobile terminal, the user does not have to pay for data transmission after downloading the game. The games are restricted by the storage and operating capacity of mobile devices.

An Example of Stand-Alone Games: Nokia Snake

Snake was the first stand-alone game that was preinstalled in Nokia's mobile handsets in 1998. Nokia owns the intellectual property rights for the application and has developed it in-house. Therefore, Nokia can install the game for free in any Nokia handset.

The idea of the original Snake was to catch more and more points with the snake steered by the player, making the snake longer and longer. At first, the player chooses the game level, which defines the speed of the snake. Finally, when the snake hits the wall or its own body, the game ends. The second version of Snak*e* was similar to the first one, but the game field had more complex shape instead of a simple box, also containing extra figures that may give extra points to the user. Both versions can be played by two simultaneous users through an infrared connection. A number of active players of the game have formed a group competing with each other in the game.

The Snake community is not as well known as the ones that the most famous console

- **Server-based games** usually require connection to the service provider's server while the game is played. The server contains the information of the game's current status.

An Example of Server-based Games: Who Wants to Be a Millionaire?™ by Codetoys

Codetoys develops mobile entertainment services for mobile operators, mobile portals, and other service providers. Codetoys' games are based on internationally recognized brands. The supported technical platforms are SMS, WAP, and iMode. One of its main products is an interactive mobile game based on the popular television show, Who Wants to Be a Millionaire?™. In addition to delivering the actual mobile game services, Codetoys provides the game platform, content, user statistics, and advice for marketing the game for mobile markets. The game can be played on all types of digital mobile phones.

Who Wants To Be A Millionaire?™ is a server-based game that is connected to the service provider's server while the game is played. The server contains information on the game's current status. The game is designed to follow the original television show concept as closely as possible. The mobile version is a multiple-choice game with a minimum of 2,000 questions. The game has 15 questions with four answers each, three lifelines, and the chance to walk away exactly as in the TV show. The object of the game is to answer 15 subsequent questions right and earn a million points. All players have a chance to make it to the hall of fame.

- **Streamed games** use advanced video decoding systems for delivering audio and visual data from servers to terminals. Streamed games require a certain minimum bandwidth for data transfer, but they will provide more advanced graphics and audio for the games on terminals that do not have the processing power for rendering demanding visual data.

An Example of Streamed Games—MatchEm TV Chat

Finland- and Hong Kong-based MatchEm Ltd. develops software systems called MiTV Tools that enable real-time interactivity between television and wireless handsets. The company's customers are TV and production companies that are willing to create interactive content into the programs delivered via analog TV, teletext, Digital TV, or the Web. MatchEm's software products are modular and can be adapted and customized for different purposes. The illustrated TV chat is one of the services based on MatchEm's iMatch product.

The vision of MatchEm is that mobile messaging should never be a stand-alone channel of communication, because the medium is too limited in its ability to deliver a robust, complete message. Rather, the medium should be used to extend the presence of a company or an event into an additional channel. Companies with a physical presence, a television, or even a Web site will be able to leverage mobile media to extend their presence to be anywhere the user is at any time.

One of the MatchEm's products is a mobile TV chat service (see figure 1), in which the end user can register a nickname and send anonymous text messages to a television show. After registration, the service confirms the availability of the name and allows sending messages to the live TV chat or to the other users. The registered users can define their profiles with user-specific information such as age and sex, which encourages the other users to contact each other. The chat service is branded as 4Date by Finnish local TV channel 4.

A Constructional View of Business Models

Rajala et al. (2001) have developed a conceptual software business model framework and practical tools for analyzing and comparing different business

Figure 1. MatchEm TV chat registration and use

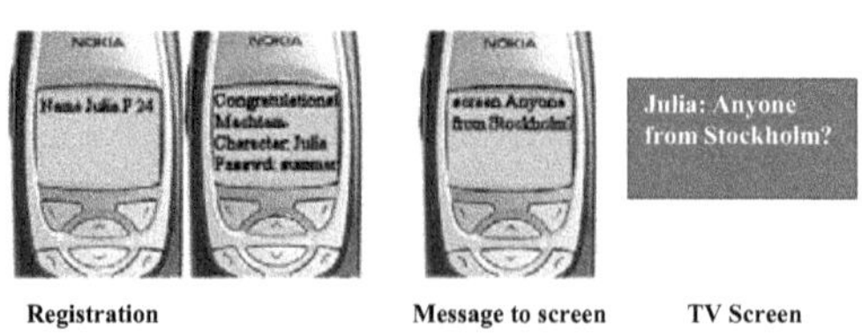

models in the software industry. We will use a subset of that framework to identify and describe the revenue logics of mobile games and related product characteristics and distribution models applied in the selected mobile game businesses. Our purpose is to identify alternative revenue logics that are technically possible, economically sustainable for various players, and that could be acceptable for the customers.

According to Rajala et al, (2001), the business model of a software vendor can be viewed as an action plan derived from strategic objectives of a company with a given product and service offering in a given market. Accordingly, a single business model deals with a single product/market situation. Consistent with the recent literature on business models (Amit & Zott 2001; Hedman & Kalling, 2003; McHugh, 1999; Morris, et al., 2004), Rajala et al. (2001, 2003) describe a business model as a combination of different functional elements of product development, revenue, sales, marketing, services, and implementation. This model includes four basic elements that can further contain several options inside them. A construction of these elements is presented in Figure 2.

In Figure 2, the business model of a software vendor is depicted by four key elements:

- The *product strategy* describes what the core product offering is and how the development of the core product of a company is organized.

Figure 2. Elements of a business model (Rajala, Rossi et al., 2001)

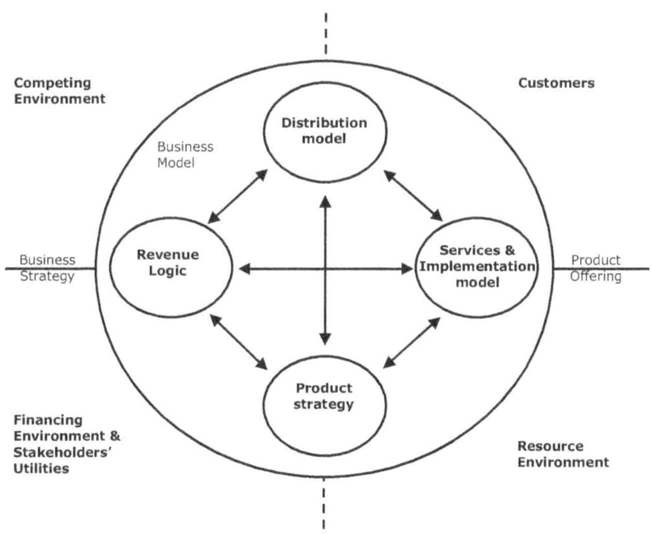

- The *revenue logic* describes how the company finances its operations; in other words, how and from what sources the revenue is generated.
- The *distribution model* describes how the marketing and sales have been organized, what are the characteristics of the channels of distribution, and who are the sellers and marketers of the product.
- The *services and implementation model* explains how the product offering is made available to the end users as a working solution.

A software company has multiple options to structure each of the elements in its business model. All of these elements are tightly interconnected with each other and cannot be analyzed in isolation. Therefore, even though our main focus here is on the revenue logic of a software company developing mobile games, we will look first at the product proposition and distribution model aspects of the business model.

Product Strategy for Mobile Games

The concepts of product strategies and product offerings are discussed widely in the literature of marketing (Cravens, 1987; Kotler et al., 1996). According to Cravens (1987), a product strategy consists of deciding how to position a product offering (e.g., specific product, line, or mix) to serve its target market, setting strategic objectives for the product offering, selecting a branding strategy, and developing and implementing strategies for new and existing products. Kotler, et al. (1996) describe a product offering on three levels. The *core product* is the essential benefit that the customer is really buying; the *actual product* includes the features, styling, quality, brand name, and packaging of the product offered for sale. Furthermore, they point out that the *augmented product* is the actual product plus the various services offered with it, such as warranty, installation, maintenance, and delivery. In the mobile game industry, these dimensions of product offerings are to be considered in both the business-to-business and business-to-consumer settings.

From the business model perspective, a defining characteristic of mobile game software as a product is that it is not a physical but an information product. Information, or digital, products have unique cost characteristics, differing largely from those of a physical product. A digital product is typically expensive to produce but very cheap to reproduce (Shapiro & Varian, 1999). In the mobile game industry, we can see that variable costs of single pieces of mobile game software are typically small, as there are no capacity constraints, and marginal costs are less than average cost. Thus, declining average costs create significant economies of scale for the producer. As the infrastructure and development tools of mobile game software evolve, the development costs of these products decrease. Simultaneously, the expected product life cycles of mobile games are sped up, and the barriers to market entry of new actors ease. In addition to having a direct effect on the game product strategies, these factors also affect the revenue logic of game software producers.

There are several dimensions in game offerings that can be used to analyze and compare different types of mobile games. First, analysis of product offerings can be made according to the intended usage scenario. This kind of analysis emphasizes the position and role of the offering in the value-creating network that produces and delivers the game offering for the end customers. According to this view, game software platforms and tools are outlined basically in different positions and roles in the industry-level value system, through which game components and final games are made available for end users. Secondly, the type and structure of the game product offering can be considered, for example, with the level of similarity of the product offering across multiple customers or customer groups and its potential distributed through different channels of distribution. This view emphasizes the potential to gain scale economies through serving a wide customer base with the same products. Third, the product development method, including various alternatives of in-house development vs. subcontracting, networking, and other forms of external development activities, can be used as a basis of industry-level classification. For example, the structure of the total offering may consist of one or more modules, including both product and service components. The structural aspect of a software product component includes the product architecture (i.e., component-based, single-core application, etc.) and the modularity in the sense of design and development. The modularity of a product potentially affects the chances for its collaborative development, including different approaches to in-house and external development.

According to the business model framework of Rajala, et al. (2003), the generic product strategy options of software vendors can be divided into five main classes, as presented in Figure 3. This classification is based on the architecture of the product offering on the level that is thought to be useful in studying different revenue logics related to specific types of product offerings.

As seen in the Figure 3, the generic options for software product offerings range from customer-specific models, where customers' needs are met with tailor-made solutions, to standardized product-oriented models, including approaches for creating universal software products and standardized online services. Between these extreme alternatives, there may be, for instance, development of parameterized system products, uniform core products, or modular product families consisting of universal software components.

Server-based and streamed mobile games typically are affiliated with online services, while stand-alone games may range from uniform core products to modular product families. In the emerging market of mobile game software, we

Figure 3. Product strategy options (Rajala et al., 2003)

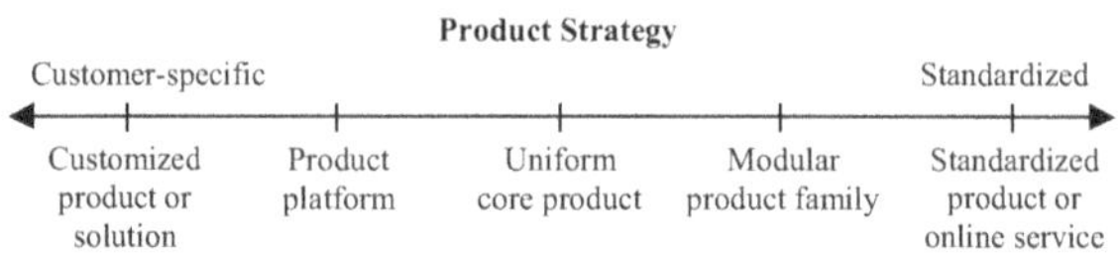

also can identify single component-based game products provided in collaboration with different partners. These components include, for example, graphics libraries or toolkits for game environments or other game components. Along with the development of the market for mobile game products, we can see an increasing diversity of game product offerings ranging from tool and platform offerings to final game solutions and product line offerings that consist of complementary games.

Distribution of Mobile Games

The development of the mobile games market has created new models for conducting business that affect all aspects of product distribution, marketing, and billing. A particularly important aspect of this new business paradigm is its impact on mobile distribution channels. Channel efficiency, channel marketing, and channel conflict are serious concerns for game developers and publishers alike.

We can identify two aspects that can serve as classification schemes in distinguishing among different distribution models of mobile game: the positioning and role of the game vendor in the mobile game value network and the complexity of the distribution system as defined by the length of the distribution channel. According to our view, the type of the distribution model and the length of the channel of distribution as one of its defining aspects strongly affect the available revenue logic options of mobile game producers.

Channels of distribution are divided in the literature, for example, into short and long channels (Lewis & Trevitt, 1996). In the mobile game business, short channels of distribution that typically consist of three or fewer than three stages of supply chain may include, for example, game developers, game publishers, and mobile operators. On the other hand, long channels of distribution may consist of more than three stages of the supply chain and typically include game developers, game publishers, aggregators, mobile portals, and mobile operators.

Short channels can offer companies possibilities to:

- Better control the sales of the product
- Monitor product sales relatively quickly and easily

Figure 4. Distribution model options (Rajala et al., 2003)

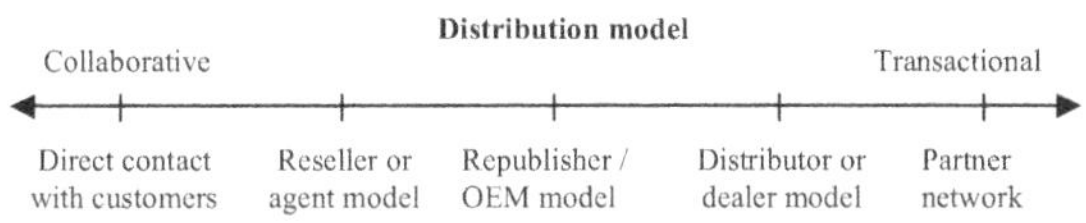

- Assist dealers' operations with advertising and promotion material
- Offer discounts and other incentives to dealers and retailers

There also are several potential disadvantages with short channels. First, the large retailers and operators will be in a better position when bargaining with the developer. Moreover, the distribution and marketing costs might increase, because the developer may have to supply to several distribution channels (Bask, 1999).

Long channels of distribution are called traditional methods of distribution channels, which are common to a wide range of products. Lewis and Trevitt (1996) identify both advantages and disadvantages with long channels related to the following themes:

- Retailers can realize all the benefits of dealing with the wholesaler
- Consumer can buy goods individually or in small quantities
- Customers usually have to pay higher prices because small retailers buy goods in bulk

There are various possibilities to reach different types of customers. For example, a firm can sell the same products through different channels, different products and services through the same channel, or different products and services through different channels.

Firms can benefit from multi-channel distribution in a variety of ways. First, it can allow them to better adapt to changing customer needs and shopping patterns. Such adaptive capability has proven useful, for example, when attempting to respond to novel distribution channels such as the Internet. Second, companies with broad product lines can benefit from a multi-channel strategy, because it is unlikely that any single channel will be optimal for all products. Third, firms with excess manufacturing capacity can benefit from additional outlets when existing channels are saturated with supply (Webb & Didow, 1997).

Use of several distribution channels and complex distribution systems offer many potential benefits to game suppliers but present some managerial challenges, as well. Multiple channels place competing demands on internal company resources such as capital, personnel, products, and technology. Moreover, the various distribution channels may compete with each other for the same customers in the marketplace, increasing the likelihood of intermediary dissatisfaction and customer confusion (Webb, 2001).

The framework of Rajala, et al. (2003) includes an aspect of the business model that deals with distributing and providing the offerings to customers. Here, the distribution model describes how the marketing and sales of the product and service offering has been organized and identifies the sellers and marketers of the product and service offering. The elementary ways of marketing a software product and service offering can been organized as illustrated in Figure 4.

In the mobile games industry, the distribution models typically include either pre-installation of the game into the mobile handset or downloadable games provided by a mobile network operator. These distribution models also are strongly tied to the pricing models. With the current GSM networks, only the smallest applications could be downloaded over the network. On the other hand, these are the only games where the network operator can act as the software distributor. Pre-installed games come as a free supplemental product. Markets are emerging for software sold separately in memory cards as commercial off-the-shelf (COTS) software.

Revenue Logic of Mobile Game Businesses

The revenue logic within a business model describes the way the software business generates revenue and profit. The different approaches to capture revenue range from different methods of pricing to different sources of revenue and different things sold. The revenue logic can include both sales revenues and other sources of financing. Here, we will focus on just the revenue element, assuming it includes the cost structure of both the offering and operation.

High initial cost and nearly zero marginal cost characterize the production and dissemination of information-intensive products (Mahadevan, 2000; Shapiro & Varian, 1999). In addition to the various revenue stream alternatives described previously, a software vendor, as in any other organization that sells electronically delivered products, has unique characteristics of the information economy to exploit. For instance, in the case of digital products, it is possible to use a range of pricing alternatives based on user segments and user-selectable options. Varian (1995) has argued that if the willingness to pay is correlated with some observable characteristics of the consumers, such as demographic profile, then it could be linked to the pricing strategy. One strategy is to bundle goods to sell to a market with heterogeneous willingness to pay (Mahadevan, 2000).

In the retail business of mobile games, a revenue model in which games are available on a subscription basis with monthly fees has met with success in Japan and for some online games services. Payment schemes allowing hardcore users to pay a flat fee for unlimited use and pay-per-use options for casual gamers make sense for maximizing volume. For the most part, only operators can employ this model at present, since they control billing for all end-user wireless services. In addition, strategies based on service subscription payment in addition to network connection charges may largely be problematic in increasing customer base.

Generic approaches to revenue logic in the software business are identified by Rajala et al. (2003) as follows:

- *Licensing*, which means license sales and royalties are the main source of revenue.
- *Revenue sharing* with distribution partners or profit sharing with users.
- *Loss-leader pricing*, which means giving something for less than its value. This is done, for example, in order to increase customer base for later revenue or to support sales of some other part of the product/service offering.
- *Media model*, where the revenue is based on advertisement sales either through advertisement in the user interfaces of software or by selling user information for advertisers.
- *Effort-, cost-, or value-based pricing* is a common approach in customized or tailor-made software solutions and made-to-order software projects.
- *Hybrid models* as various combinations of the previous points.

In the following paragraphs, we discuss selected approaches to potential revenue logics in the mobile game business.

Licensing

Licensing is the most common revenue model in the mobile industry (Durlacher Research Ltd., 2001). This revenue model is identified by Hecker (1999) as being a part of the standard software business model. It involves selling the customer the right to use the software. In licensing, there are many alternatives, including per-user, per-machine, per-concurrent user, or site licensing. Revenue structure may include some amount of upfront payment for the integration of the wireless solution itself and revolving license payments over the life cycle of the contract. This may depend on the number of the users or number of applications. Unfortunately for application providers, network operators often retain a major share of revenues.

Revenue Sharing

Revenue sharing is a common practice between partners in the channels of distribution (e.g., between game developer and mobile operator). Instead, profit sharing is usually limited to B-to-B settings only between the user and producer of a piece of software. Profit sharing is essentially a form of licensing in the sense that it also involves selling the right to use the software. However, in this model, the software provider's revenue is tied to its customer's performance when using the software.

A logical choice of a model for wireless games companies is based on revenue sharing with network operators, who provide the backbone for transmission game data. Empirical observations indicate that network operators often retain a major share of the revenue. However, in some cases, the service provider may charge the end user directly without an operator taking their share of the data transfer revenue.

Loss Leader Model

The loss leader pricing model here means giving something for less than its value. This is done in order to increase the customer base for later revenue or to support sales of some other part of the product/service offering. An example of loss leader revenue logic is a model in which the software is provided for free, and revenue is collected through selling related products or services to the users, or from the sales of complementary offerings to other customers. Hecker (1999) introduces the term *support selling* to illustrate cases in which revenue is collected through media distribution, branding, training, consulting, custom development, or after- sales support. Glynn (1999) notes that offerings provided for free are not merely an incentive, but, ideally, they also stimulate the usage of fee services. Examples of this approach can be seen in games that are preinstalled in mobile devices. These games can be developed in-house by the handset manufacturer or outsourced to a third-party software company.

Media Model

Hagel and Armstrong (1997) point out that the media revenue model, for instance, is an essential part of the virtual community business model. In a media model, the software is used to collect a group of users. For example, access to this group of users may be sold to third parties for advertising purposes. The media model involves a multitude of arrangements, in which third parties can be provided with information about the users, users are provided with information about the services of third parties, and the software acts as the mediator. This approach will produce interesting revenue opportunities in the future, as the user segments of mobile games make interesting target groups for many advertisers, and they may share some preferences or demographical and cultural characteristics.

Summary and Conclusion

In this chapter, we have focused on channel choices and revenue logics of mobile game producers. Using parts of the business model framework developed by Rajala et al. (2003), we have discussed potential revenue models of mobile game developers and identified examples of these in the mobile game industry.

In the existing and potential delivery models of mobile games, revenue can be collected either from user licenses, from royalties based on sold copies, from transactions concluded while the game is played, or from both. Furthermore, the possibility to deliver mobile games online offers a way to reach a large number of users. However, this possibility strengthens the role of mobile operators as distributors in the mobile gaming businesses, because the delivery and billing processes play a key role in the revenue logic. It seems, therefore, that the current the winners in the industry are the telecom operators, who dominate the end-user interface. They have close relationships with the users and viable delivery and billing mechanisms. However, they have not yet been able to create a critical mass of users or the volume needed to convert mobile entertainment into a profitable business. These circumstances create opportunities for other players with new business models.

The expected growth of the entertainment business combined with increased mobility offer a number of opportunities even for small software companies that develop mobile games. Research of viable business models for these companies is highly necessary. There is an emergent need to identify and analyze the success factors and key characteristics of business models of companies developing mobile games. This will improve the understanding of business models as well as provide valuable information for the companies involved in the mobile entertainment services industry.

References

Amit, R., & Zott, C. (2001). Value creation in e-business. *Strategic Management Journal, 22*(6/7), 493-520.

Bask, A.H. (1999). *Third party relationship in logistics services*. Helsinki: Helsinki School of Economics.

Cravens, D.W. (1987). *Strategic marketing*. Homewood, IL: Richard D. Irwin, Inc.

Durlacher Research Ltd, E.P.O. (2001). *UMTS report: An investment perspective*. Durlacher Research Ltd.

Glynn, S. (1999). Making money from free services. Mercer Management Consulting.

Hagel, J.I., & Armstrong, A.G. (1997). *Net gain—Expanding markets through virtual communities*. Boston, MA: Harvard Business School Press.

Hecker, F. (1999). *Setting up shop: The business of open-source software*. Open Resources.

Hedman, J., & Kalling, T. (2003). The business model concept: Theoretical underpinnings and empirical illustrations. *European Journal of Information Systems, 12*, 49-59.

Kotler, P., Armstrong G., Saunders, J., & Wong, V. (1996). *Principles of marketing*. Hertfordshire, UK: Prentice Hall.

Lewis, R., & Trevitt, R. (1996). *Intermediate retail & distribution*. London: Hodder and Stoughton.

Mahadevan, B. (2000). Business models for Internet-based e-commerce: An anatomy. *California Management Review, 42*(4), 55-69.

McHugh, P. (1999). Making it big in software—A guide to success for software vendors with growth ambitions. Rubic Publishing.

Morris, M.,. Schindehutte, M., & Allen, J. (2004). The entrepreneur's business model: Toward a unified perspective. *Journal of Business Research* (forthcoming).

Rajala, R., Rossi, M., & Tuunainen, V.K. (2003). A framework for analyzing software business models. *Proceedings of the 11th European Conference on Information Systems*, Naples, Italy.

Rajala, R., Rossi, M., Tuunainen, V.K., & Korri, S. (2001). *Software business models—A framework for analyzing software industry*. Helsinki: The National Technology Agency of Finland.

Shapiro, C., & Varian, H.R. (1999). *Information rules, a strategic guide to the network economy*. Boston: Harvard Business School Press.

Varian, H.R. (1995). *Pricing information goods*. MI: University of Michigan.

Webb, K.L. (2001). Managing channels of distribution in the age of electronic commerce. *Industrial Marketing Management, 31*, 95-102.

Webb, K.L., &. Didow, N.M. (1997). Understanding hybrid channel conflict: A conceptual model and propositions for research. *Journal of Business-to-Business Marketing, 4*, 39-78.

Chapter XVI

The Role of Business Models in Developing Business Networks

Jukka Heikkilä, University of Jyväskylä, Finland

Marikka Heikkilä, University of Jyväskylä, Finland

Markku Tinnilä, Helsinki School of Economics, Finland

Abstract

Business models have received a substantial amount of interest recently. Also, various research studies have discussed business models, especially in the context of a single company operating in mass markets. Unfortunately, these models often are not applicable for complex products or services that build on long-term knowledge about customer tastes, facilities, and skills. Especially on global markets, the asset specificity and vast geographical distances make it difficult for any single company to provide this kind of service cost-efficiently on a large scale. Instead, it calls for cooperation among multiple firms. Creation of a joint business model for a collaborative network is a necessary means by which companies can coordinate cooperation in practice. The CSOFT metamodel proposed in this chapter provides guidance for joint business model development by emphasizing customer relationship and adjustment processes needed within the network.

Introduction

Business models have recently been a hot topic. Since the end of 1990s, there has been a vivid research stream proposing differing definitions, lists of components, taxonomies, change methodologies and evaluation models for business models (Bouwman, 2003; eFactors, 2002; Faber et al., 2003; Osterwalder & Pigneur, 2002). In essence, the topics discussed in the business model literature are not new; the components of business models have been recognized, at least to some extent, in business strategies and business planning for decades. But the need for explicit analysis and description of the business model has become more inevitable as the introduction of information and communication technology has enabled completely new ways of making business.

To synthesize, a business model to tell us how strategy is implemented by describing, for example, its product, infrastructure, financials, technology, and customers and their relationships: "A business model depicts the content, structure, and governance of transactions designed so as to create value through the exploitation of business opportunities" (Amit & Zott, 2001, p. 493). In addition, this implementation is related to its upstream and downstream business environments.

Most often, business networks are formed in order to make the existing supply chains more efficient (Van de Ven, 1976). However, the networks between partners and even competitors also can be seen as not only a means of cutting costs, but also as a source for new innovative business ideas, where the network provides the customer with more added value than if the companies were operating independently (Nalebuff & Brandenburger, 1996).

In these business networks, independent companies form consortiums or alliances in order to jointly serve customers with new offering and even to jointly develop the underlying business models. This calls for reconsideration of each component of the business model, especially the revenue sharing and roles of each participant. The new business models, therefore, must describe in sufficient detail the value created to the customer and the income generated to each participant. In addition, each partner's role in producing the output has to be explicit.

The most prominent examples are found in mobile and Internet businesses, where telecom operators aim at producing by themselves a very wide range of services and content. However, the operators soon find out that they cannot compete with specialized and established media houses in content production. The media houses have similar plans to find a direct access for their content in the Internet by circumventing the middlemen. In both cases, there have been a number of failed business models. The success stories, however, have been networks of operators and content/service providers, where some of the parties moved to the middle of their value chain (Clemons et al., 1993). For example, the Japanese operator, NTT DoCoMo's i-mode product successfully mediates the output of a network of content providers to customers and serves as a good example of the possibilities of win-win business models in networks (Saarinen et al., 2002).

Building upon Faber et al. (2003) and Osterwalder and Pigneur (2002), we propose a metamodel for joint business model development that differs from the previously mentioned models in that it considers joint produced services that require long-term relationships. This puts an emphasis on the customer relationship and adjustment processes needed within the network to enable the joint service concept in the first place. The metamodel starts with the definition of ontology, defining the most important factors in the joint business model (i.e., customer relationship, service, organization of the network, finance, and technology). It then guides the analysis, negotiations, and change processes needed and points out customer- and case-specific limitations affecting the business model. Finally, by taking into account the limitations of the network, we end up with a feasible, networked business model.

The Traditional View of Business Models

Traditionally, a company's business is built on strategy, which is typically reflected at operations-level business plans for specific products or product groups. However, there often is quite a substantial gap between these two levels. In particular, today's rapid technological change, along with the societal, organizational, and individual factors in the global market, makes the management and planning of new business ideas complex and challenging (eFactors, 2002) (see Figure 1). Therefore, there is a need for a systematic description of a business model, methods, and processes for turning the strategic visions into organizational arrangements, responsibilities and roles, information, material and money flows, and so forth. Even though these aspects probably have been included in well-prepared business plans, the virtue of business models is that they consider the business context from a conceptual level, which makes it somewhat independent of current processes and restrictions of the company. Instead, a business model may be used as a tool (i.e., a conscription device while developing the model) and later on as a boundary object (Boland & Tenkasi, 1995; Brown & Duguid, 1991) to point out the need for additional capabilities and resources, and to figure out alternative ways to organize transactions and use new technologies (Heikkilä et al., 2004a).

Figure 1. Business model and context (synthesis from Osterwalder & Pigneur, 2002 and e-factors, 2002)

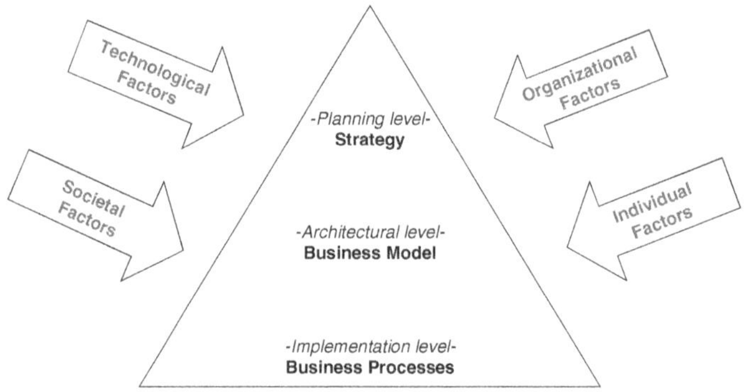

The business model can be seen "as the strategy's implementation into a conceptual blueprint of the company's money earning logic" (Osterwalder, 2004, p. 14). In other words, the vision of the company and its strategy are translated into value propositions, customer relations, value networks, technology, and financial arrangements. The business processes level, in turn, is the materialized form of the conceptual business model, appearing as workflows and so on.

Similarly, adoption of new strategies may require reengineering of business processes and IT. Several studies have noted the relationship between successful implementation of strategic change and the need to change business processes (Kallio et al., 1999). Our discussion builds upon recent literature on business models (eFactors; 2002; Faber et al, 2003; Osterwalder, 2004; Osterwalder & Pigneur, 2002), on research on organizations' capabilities of managing alliances (Andersen & Christensen, 2000; Powell, 2000), and on the capabilities to change in their network relationships and infrastructure (Kumar & van Dissel, 1996; Möller et al., 2004; Ring & Van de Ven, 1994; Van de Ven, 1976). We present a new framework that takes into account the effects of networks.

Joint Business Models: The Importance of Networks

Most business model literature has focused on analyzing models for a single firm or organization (Bouwman, 2003). However, in recent years, various kinds of networks have become common in business. For example, studies by Tsupari et al. (2004) show that over 80% of companies in the manufacturing industry in Finland are involved in networks to some extent. Furthermore, the importance of networks most probably will be even higher in near future, as the tendency seems to be toward deeper relationships requiring higher levels of commitment and coverage. Outsourcing of non-core functions of companies amplifies this trend and increases the need for managing networks. At first, the relationships most often are traditional buyer-seller ones but evolve over time to closer partnerships and network relations. Thus, in reality, the business model analysis of a single company reveals usually only a small part of a complicated service network.

The relationship of the development of a joint business model with the strategy implementation process can be described as follows (Heikkilä et al., 2004a). Typically, a company has an initial idea or a business problem regarding further processing that requires the knowledge and core competence of several firms. The very first task in the formation of a network is the initial negotiation phase, which concerns communication and understanding the ambitions of the potential parties and the selection of a right or suitable combination of capabilities and resources. All potential partners have to make their own decisions as to whether or not to engage in the network and then decide which business units should be involved.

Only after the core companies of the network have joined can the actual negotiations over specific goals (i.e., inter-organizational strategy adjustment) be started. Each potential participant of a business network has its own strategies, models, and processes for the present and the future. Therefore, the network should create a joint business logic that matches or complements each company's strategic objectives. This means that each partner in the network should reveal its true strategic goals concerning its cooperation, after which the network may jointly make decisions over the target for the network. Without a doubt, this task is a demanding one, and it often takes a considerable amount of time to build a sufficient level of trust between the parties before strategic intentions are articulated and communicated and actions are taken accordingly.

Before this trust is achieved, there may be several obstacles. For example, different units of network members may not share a common view of the benefits of joining the network. Furthermore, the strategic advisability for partnering may be marred by short-term needs to generate income. Indeed, the widespread adoption of short-term management through increasing shareholder value may be a major stumbling block in the road of many networks.

Based on a case study (Heikkilä et al., 2004c), we noted that if the interorganizational strategic adjustment is carried out as a Scandinavian style discussing process, it might become a never-ending story. Whenever a new member, either individual or company, enters the consortium, new negotiation and sense-making rounds are restarted. At some point, it becomes necessary to choose a leader or to agree on the focal company that should take the responsibility for coordinating the cooperation. The focal company is most naturally the one providing the most critical core competence in the new service concept, as it has the greatest motivation to create a new business model.

If strategic intentions are aligned to a sufficient degree, then the network can start developing the joint business model. Perhaps the most beneficial way is to formulate the model simultaneously with top-down and bottom-up approaches. When looking at a business model from a top-down view, the network aims at pointing out necessary conditions for joint business. In the bottom-up approach, the requirements arising from practical business processes are reflected in the very same business model, calling for changes both in intraorganizational processes and in the espoused strategies of each participating company.

Gradually, when the business model is shaping up, the focus moves more toward practical business processes needed to implement the business model. Evidently, interorganizational process adjustment should take place. This means that the members should look at the kinds of processes they already have and how the network could, by combining these processes, produce the desired outcome. Simultaneously with the above-mentioned adjustment, each member should reflect on its findings from the benefits of cooperation to its own internal processes and strategies and make adjustments.

To summarize, the participating companies should engage initially in three adjustment processes: horizontally at the strategy-business model interface between the companies (Powell, 1990); horizontally at the processes-business model interface between the companies; and vertically within each company to align the strategies and processes to meet the challenges of cooperation (e.g., Takeishi's internal coordination of interfirm cooperation) (Takeishi, 2001). There is also an evident need for a fourth adjustment; namely, to find out the uncovered parts of the business model, when needed.

A Metamodel for Joint Business Models Development in Practice

The process of developing of a joint business model can be an esoteric exercise of intellect and imagination. However, the dirty details of the real world have to be taken into account, if the business model is ever to be implemented, especially in a cost-efficient manner. For example, a single organization may need more than one business model, because it may have multiple business sectors and products or services. These several business models then require different organizational arrangements and ICT support, especially with different product/market combinations, as argued previously. This means that a business model

Figure 2. The metamodel

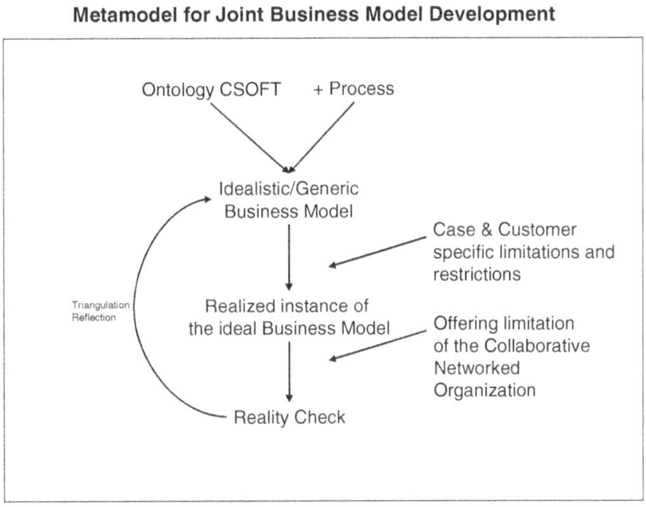

should have the general characteristics of a good model; it should describe the ontology, process, and means for sensitivity analysis for the designer. The procedure with which to identify a good model is depicted in Figure 2.

The procedure for developing a joint business model starts with the ontology. Ontology is an explicit simplified conceptualization of the objects, concepts, and other entities that are assumed to exist in some area of interest—in this case, joint business models—and the relationships that hold among them. In addition to the ontology, creating joint business models includes definitions of the change processes needed. Then, the procedure continues with recognition of demand-side restrictions. These determine the realized instances of a business model. The last step is to consider supply-side restrictions, which may affect the possibilities for the network to offer specific products or services as planned. This requires a reality check or proofing of the concept (i.e., business model should be critically analyzed and fed back to the previous stages) (Heikkilä et al., 2004b). Only after passing this reality check can the collaborative network start sales negotiations with potential customers.

Next, we describe the CSOFT Ontology, consisting of customer relations, service, organization of network, finance and technology. Then, we define case- and customer-specific limitations, offering restrictions, and their impacts on business models.

Figure 3. CSOFT ontology

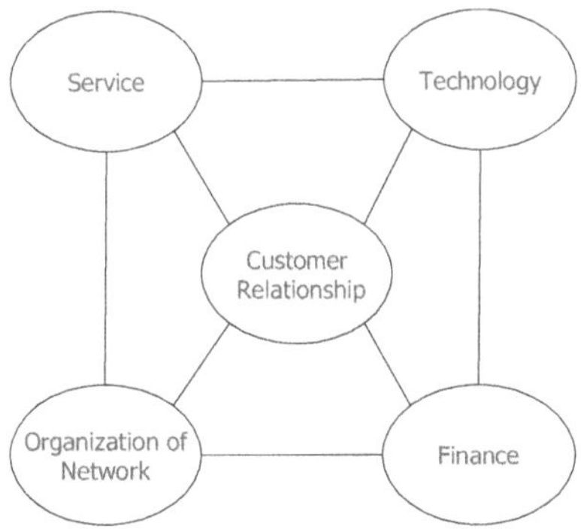

Ontology: Five Factors of CSOFT

Whereas the models by Osterwalder and Pigneur (2002) and Faber et al. (2003) that inspired us primarily focused on short-term transactions in the mass market, our emphasis is on long-term service models for B-to-B markets. It can best be characterized by the long-lasting business relationships among customers and suppliers, mutual interests of the consortium companies in serving the complex technical infrastructure, and their interdependency in providing global service. In this context it is the customer relationship that deserves special attention. It is the customer relationship and its connections to other parts of the metamodel that make devising the business model in this context so complex.

We developed a CSOFT ontology that defines the business model as a combination of several interrelated elements. The central distinguishing element is the nature of the customer relationship, which, in our context, plays a major role but is neglected in many other business models. It is connected to other factors such as service, organization of network, finance and technology, as depicted in Figure 3.

Customer Relationship

The customer relationship is the focal point in CSOFT ontology. In B-to-B business, the relationships are often very long. In our case (Heikkila et al., 2004a, 2004b, 2004c), it is typical to have well over half-a-century-long relationships with customers. This brings up the question of relationship ownership—who owns the customer in a networked business model? No straightforward answer can be given, but clearly it depends on the service offered to customers. In jointly developed products/services or bundled existing products, the companies in the network should jointly own the relationship.

There are different approaches to customer relationships. In CRM, the approach is rather mechanical (i.e., use of IT in the management of relationships is emphasized). The origins of CRM are in the B-to-C business, where the use of frequent customer programs, for example, are common. Most airlines (e.g., Finnair) and retail chains (e.g., SOK, Kesko) run similar types of customer programs. However, the number of customers is lower in the B-to-B business, which is reflected in the customer relationship management in industrial relations. Usually, the relationships are managed by key accounts or product managers, who frequently making personal contacts with customers. Large IT providers, such as HP or IBM, are prime examples of this. However, the methods used in CRM are useful in industrial relations, too.

Service Component

The service component depicts the intended value of the service and how it is created and provided. This includes defining the type, or different types, of services provided, as well as the service and production processes behind the service.

The service provided to customers is the key value-creating element in the business model. The type of service provided may be organized in several ways. Increasingly, industrial enterprises have separate service processes for different products and services. Kone Elevators, Kone Cranes, and several other companies have made a difference in providing standardized, modularized, and customized products by using differentiated processes. This also applies to renewal and service processes, where similar structures can be found.

To create new business models, there is usually a need to redesign processes in a true reengineering sense. Often, the redesign of processes is dependent on new applications of ICT technology; for example, many industrial services (SKF) or extended services (Würth) providing value to customers.

Organization of Network Component

The organization of network component defines the roles of networks participants (i.e., the tasks and operations that are performed by each partner). This division of tasks is not stable over time and is subject to changes due to the development of processes and services, as well as changes in customer relationships and the entry of new network partners.

Some networks are very dynamic by nature, resembling a project organization, where the organization is formed individually for each customer and task. This is typical in large-scale construction projects, airline production, and paper machine deliveries. In some networks, there is a clear subcontracting relationship with one focal company, as in the so-called Nokia cluster. The network also may be based on operational partnerships, where network members produce services and content dedicated to end-users via one partner. This is the case of the Japanese i-mode service, where NTT DoCoMo is the key player. Other operator-driven networks, such as Vodafone live!, are similar. Strategic alliances are relationships with equal partners, where the long-term relationship is emphasized. The car industry has several examples of this (e.g., Ford-Mazda relationship).

Then, how dynamic or stable should a network be? If we bear in mind the half-a-century life cycle, then this would call for a network where at least its core would be stable, even though the outer rim of the network would be more dynamic, correlating, for instance, with changes in customer or market segments.

Finance

The finance component focuses on cost issues and revenue sharing. These are major stumbling blocks in many networks. There are multiple examples of this in e-business, where at one time it seemed that everyone was willing to accept partners in their network on the condition that the major share of revenue generated would end up into their pockets. Needless to say, most of these unequal networks failed even before launch. Examples include multimedia services such as WAP (Wireless Application Protocol platform) and Internet portals. The network did not have a joint idea of a fair sharing of revenues, and, as a consequence, the members (i.e., operators, contents producers, media houses, IT developers, etc.) all felt they received an inadequate share of the revenues generated by the networked service.

New financing and pricing models often form a vital part of new services. The trends have originated from office equipment services, where copy machines by Xerox have been leased with full service support for decades. Similarly, car leasing services offer extension to old services with a financing element, relieving the companies from investing. This new trend of new service (business) models based on financing now includes such examples as Dell in leasing and maintenance of computers, Securitas in offering security services bundled with rented equipment by a monthly payment, and construction companies in offering partial ownerships. There are several different models available, but few of them are used presently in complex industrial facilities.

Technology

Technology for providing services includes especially ICT support for the business models. As most businesses nowadays rely heavily on ICT for operations, a proper ICT infrastructure is essential for success. One of the main obstacles in technology is the incompatibility of the systems of network partners for exchanging data/information and education materials and training services. If a new business model needs a complete revision of ICT, the cost may prove to be a barrier of starting operations, as in the case of many banks being slow to offer Internet banking services.

Change Processes

As discussed earlier, there are multiple adaptation processes needed for creating and implementing a joint business model. There is (1) adaptation at the strategy level of the companies; (2) adaptation at the business practices, infrastructure, and tools; and (3) their alignment to meet the challenges of cooperation (Heikkilä et al., 2004a). This means that the actual business model should include explicitly the processes to pay attention to in these adaptation processes. We suggest four types of change processes during the design stage:

- A process guiding the analysis of customer characteristics and case-specific needs (i.e., facilities) and negotiations of the roles and offering of the consortium, because it can be very time consuming.
- A process guiding the implementation of the business model. Within the consortium, this means the three previously mentioned adaptation processes for matching strategic intentions, internal change processes, and interorganizational changes in processes and infrastructure. For a customer, these changes should provide simplicity and benefits.
- A process guiding the identification and follow-up of real costs of producing the service and transaction costs (Williamson, 1985). This may mean, for example, creating proofs of concepts, service level agreements, quality of service, intellectual property rights ownership, and the ownership of assets in advance on a contractual basis.
- Finally, a process of devising a road map for the firms to change and of getting the necessary people involved on a regular basis in evaluating the viability of business models from different perspectives (Hoffner et al., 2004).

Demand-Side Limitations and Realized Instances of the Ideal Business Model

By describing the ontology and the change processes proposed, we can generate an idealistic business model. However, in reality there are always some demand-side limitations that must be taken into consideration in the business model. Thus, in the next phase in the creation of a joint business model, the companies should find out case- and customer-specific limitations and restrictions and analyze their effects on the feasible realized business model instances. In practice, this means that if the markets are not homogenous, either the customer segment is reduced in size, or there most likely will be several different realized instances of the business model.

Supply-Side Limitations and Reality Check

Similarly, the supply side also has some limitations (limitations in the offering of a cooperative network) that affect the capability to perform. For instance, if the network is lacking expertise needed for a specific product or service, it either has to rule this service out from its offering, or it must change the composition of the network. As suggested earlier in this chapter, in terms of offering and organization of the network, this implies that various market segments or areas should be served partially by different networks. Finally, a reality check for the business model can be carried out with tentative proof of concepts, prototypes, and benchmarking.

Conclusion

In this study, we analyze the development of a networked business model in a context where a number of companies are to develop novel businesses jointly. Previous studies on business models primarily cover business models of one organization for the mass market. But imagine a situation where you are asked to service, repair, or upgrade a physical production system that was built years ago in an overseas market. You are asked to do that because you originally delivered some parts to the system, and, at present, you are one of the leading suppliers of equivalent facilities. The problem is that the facility has been changed and even become outdated, but it is still to be kept running for several years. The potential business relationship urges you to work in a more clever way than before, utilize a number of partners and local service providers, and do all this in a cost-efficient manner on a global niche market. This calls for a novel joint business model.

Organizations that participate in such a network naturally have each their own business strategies for the present and the future. If the cooperation is to succeed, the companies should adjust their operations and even strategies to some extent. As a consequence, the joint network emerges incrementally through several adjustment processes. The CSOFT metamodel presented in this chapter can be regarded as a means of coordinating these several cyclical adjustment processes needed in an effort to state the joint offering of the network. Thus, in the sense of Argyris and Schön (1978), it is a device enabling higher-level organizational learning. It is especially designed for networks offering complex services. Because of the context, it emphasizes the importance of a customer. This metamodel provides not only the elements, but also guides the process of creating, analyzing, and evaluating potential business models for a startup network. Moreover, it is planned to be applicable each time the network designs new offerings for its customers.

Note

This research is financially supported by the Academy of Finland, National Technology Agency of Finland, and the participating companies.

References

Amit, R., & Zott, C. (2001). Value creation in e-business. *Strategic Management Journal, 22*, 493-520.

Andersen, P., & Christensen, P. (2000). Inter-partner learning in global supply chains: Lessons from NOVO Nordisk. *European Journal of Purchasing & Supply Management, 6*, 105-116.

Argyris, C., & Schön, D. (1978). *Organizational learning: A theory of action perspective*. Reading, MA: Addison Wesley.

Boland, R., & Tenkasi, R. (1995). Perspective making and perspective taking in communities of knowing. *Organization Science, 6*(4), 350-372.

Bouwman, H. (2003). State of the art on business models, B4U/D3.2. Telematica. Instituut, TU Delft, TNO, KPN research, 42.

Brown, J., & Duguid, P. (1991). Organizational learning and communities-of-practice: Toward a unified view of working, learning and innovation. *Organization Science, 2*(1).

Clemons, E., Reddi, S., & Row, M. (1993). The impact of information technology on the organization of economic activity: The "move to the middle" hypothesis. *Journal of Management Information Systems, 93*(10), 2, 9-35.

DiMaggio, P. (Ed.). (2003). *The twenty-first-century firm: Changing economic organization in international perspective.* NJ: Princeton University Press.

eFactors. (2002). WP3: E-business model roadmap. Deliverable 3.1. of Information Systems Technologies e-Factors Thematic Network Report, IST-2001-24868.

Faber, E., et al. (2003). Designing business models for mobile ICT services. *Proceedings of the 16th Bled Electronic Commerce Conference*, Slovenia.

Heikkilä, J., Heikkilä, M., & Lehmonen, J. (2004a), Joint development of novel business models. *Proceedings of the IFIP 18th World Computer Congress, 4th International Conference on I3E,*

Heikkilä, J., Heikkilä, M., & Lehmonen, J. (2004b). Sharing ror understanding and doing for learning: An emerging learning business network. *Proceedings of the 27th Information Systems Research Seminar in Scandinavia,* Falkenberg, Sweden.

Heikkilä, J., Heikkilä, M., Lehmonen, J., & Pekkola, S. (2004c). Smart ICT support for business networks. In P. Vervest, E. Heck, K. Preiss, & L.-F. Pau (Eds.), *Smart business networks* (pp. 389-404). Springer.

Hoffner, Y., Field, S., & Facciorusso, C. (2004). The typed domain—A recipe for creating virtual enterprises. *Proceedings of the IFIP WCC 2004.*

Kallio, J., Saarinen, T., Salo, S., Tinnilä, M., & Vepsäläinen, A.P.J. (1999). Drivers and tracers of business process changes. *Journal of Strategic Information Systems, 8*(2), 123-140.

Kumar, K., & van Dissel, H.G. (1996). Sustainable collaboration: Managing conflict and cooperation in inter-organizational systems. *MIS-Quarterly, 20*(3), 279-300.

Möller, K., Rajala, A., & Svahn, S. (2004). Strategic business nets—Their types and management. *Journal of Business Research* [in press].

Nalebuff, B., & Brandenburger, A. (1996). *Co-opetition.* Profile Books.

Osterwalder, A. (2004). *The business model ontology—A proposition in a design science approach* [doctoral thesis]. University of Lausanne.

Osterwalder, A., & Pigneur, Y. (2002). An e-business model ontology for modelling e-business. *Proceedings of the 15th Bled Conference on E-Commerce*, Bled, Slovenia.

Powell, W.W. (1990). Neither markets nor hierarchy: Network forms of organization. *Research in Organizational Behavior, 12*, 295-336.

Powell, W.W. (2000). The capitalist firm in the 21st century: Emerging patterns.

Ring, P., & Van de Ven, A. (1994). Developmental processes of cooperative interorganizational relationships. *Academy of Management Review, 19*(1), 90-118.

Saarinen, T., Kallio, J., Tinnilä, M., & Vesa, J. (2002). Customer relationship management in service mediary driven mobile services: Case i-mode. In T. Reponen (Ed), *Information technology enabled global customer service*. Hershey, PA: Idea Group Publishing.

Takeishi, A. (2001). Bridging inter- and intrafirm boundaries: Management of supplier involvement in automobile product development. *Strategic Management Journal, 22*, 403-433.

Tsupari, P., Sisto, J., Godenhjelm, P., Oksanen, O.-P., & Urrila, P. (2004). Yritysten liiketoimintasuhteet. Selvitys liiketoimintasuhteista ja verkostoitumisesta suomessa (business relationships between companies. A study on business relationships and networks in Finland), Katsauksia 2004/6. Tilastokeskus, 78.

Van de Ven, A.H. (1976). On the nature, formation, and maintenance of relations among organizations. *Academy of Management Review*, 24-34.

Williamson, O.E. (1985). *The economic institutions of capitalism; firms, markets, relational contracting*. New York: The Free Press.

Chapter XVII

Are Multi-Channel Technologies Adopted in Online Consumer Auction Markets in Finland?

Jarkko Vesa, Helsinki School of Economics, Finland

Eric van Heck, Erasmus University's Rotterdam School of Management, The Netherlands

Abstract

In this chapter, we analyze factors relevant for adopting multi-channel technologies in online auction markets. Its focus is on how multi-channel technologies, such as Web and Internet technologies, digital TV technologies, and wireless technologies, are used in online consumer auction markets. Using a conceptual framework that identifies four factors that could explain the adoption of multi-channel technologies, we analyze five Finnish consumer auction markets: Huuto.Net, Keltainen Pörssi, QXL Finland, Systeemi.Net, and Tori. Our analysis shows that the Internet and World Wide Web are the predominant—and, in most cases, the only—online channel available to the users. Furthermore, not even the Web technologies are used to their full potential. This chapter suggests that media richness and the ability to provide multiple modes of communication relationships stimulate the adoption of multi-channel technologies. Conclusions are presented and implications are drawn for future research.

Introduction

Since the beginning of the 1990s, various types of online markets have become increasingly popular (Kambil & van Heck, 2002; Lucking-Reiley, 2000). One of the key drivers behind the increase has been the rapid proliferation of the use of the Internet both by consumers and by firms. Various types of closed information systems had been used in the implementation of online markets before the Internet became widely accepted, but as Malone (1995) has argued, the market prefers the decentralized and open Web for electronic commerce to the traditional, centralized, and closed environments provided by different types of service providers. From the online auctioneers' point of view, the Internet makes auctions accessible even to novice bidders. In other words, one doesn't have to be an expert in auctions in order to participate in online auctions. Another nice feature of an auction in an online environment is that consumers can participate simultaneously in several auctions, giving them freedom to surf among Web-based auctions all around the world (Bapna et al., 2003).

There also are two other trends (including online auctions) that make online commerce more attractive for consumers. First, the number of retailers offering multiple channels (i.e., traditional brick-and-mortar, mail-order catalogs, and virtual online and wireless channels) is increasing; and second, as a result of rapid technological development, the Internet has become a powerful and efficient search-purchase tool (Shim et al., 2004). We argue that online markets represent a novel search-purchase tool that consumers will adapt to at an increasing rate in the future, as the success of eBay suggests.

Therefore, we can anticipate that, along with the growth of the Internet, both the number of consumers entering the world of online auctions and the number of auctions they participate in will grow, provided that online auctioneers manage to implement exchange processes and supporting technologies in such a manner that consumers not only try their services once, but continue to participate in online auctions again and again.

In this chapter, we will analyze the relevant factors for adopting multi-channel technologies in online auction markets by focusing on how multi-channel technologies (e.g., Web and Internet technologies, digital TV technologies, wireless technologies) are used in online consumer auction markets. Then, using a conceptual framework that identifies four factors that could explain the adoption of multi-channel technologies, we will analyze five Finnish consumer auction markets. The reason for choosing online auctions and, in particular, consumer-to-consumer online auctions as the context of our research is related to the fact that the exchange processes of auctions typically are advanced by nature; thus, the focus of this explorative research is on the relationship between multi-channel technologies and the exchange processes of online markets and not on online auctions as such.

The structure of this chapter is as follows. We will begin by reviewing online markets and exchange processes in the next section. In following section, we will present propositions derived from the review and the conceptual framework used in our analysis of the relationship between relevant factors for the adoption of multi-channel technologies. Section 4 presents the five Finnish online auction markets and the methodology used in this chapter. The next section presents and discusses the results of our empirical analysis, and the final section provides our conclusions.

Online Markets and Auctions

According to Bakos (1998), both traditional and electronic markets have three main functions: "matching buyers and sellers; facilitating the exchange of information, goods, services and payments associated with market transactions; and providing an institutional infrastructure, such as a legal and regulatory framework, that enables the efficient functioning of the market" (p. 35). He explains that the first two functions (i.e., matching buyers and sellers, facilitation of transactions) are typically provided by intermediaries, while the government is providing the institutional infrastructure upon which the intermediaries build their operations.

An electronic market allows buyers and sellers to exchange information about prices and product offerings by using information technologies (Bakos, 1991). Bakos (1998) argues that emerging, Internet-based electronic markets "leverage information technology to match buyers and sellers with increased effectiveness and lower transactions costs," and therefore lead to "more efficient, 'friction-free' markets" (p. 35). As a result of the increased use of electronic communications and the Internet, in particular, "the costs of searching for trading partners, verifying their capabilities, and monitoring contracts fall dramatically" (Kambil & van Heck, 2002, pp. 18-19). This, according to Malone, et al. (1987) should lead to an increased use of markets and outsourcing. We can conclude that electronic markets, or online markets, will provide new business opportunities both for business-to-business and business-to-computer markets.

Online auctions, also known as electronic auction, or e-auctions, have been studied extensively in recent years (Bakos, 1998; Kambil & van Heck, 2002; Lucking-Reiley, 2000; McDonald & Slawson Jr., 2002; Paarlberg, 2001; Strader & Ramaswami, 2002). The main driver for the interest appears to be the huge success of eBay, the leading consumer-to-consumer online auction in the world. Finnish online auctions also have been subject to academic research (Laine, 2002; Puhakainen & Tuunainen, 2001; Vesa & van Heck, 2003).

There are three basic types of Internet-based auctions: business-to-consumer (B2C), consumer-to-consumer (C2C), and business-to-business (B2B) online auctions. According to Bapna et al. (2003), these auctions represent a new class of mercantile processes that are ushering in the networked economy. In this study, we will focus exclusively on consumer-to-consumer online auctions. The number of consumer-oriented online auctions (i.e., B2C and C2C auctions) is huge. Paarlberg (2001) reviewed 196 online auctions in eight European countries; Lucking-Reiley (2000) included 142 different auction sites in his study; and Puhakainen and Tuunainen (2001) identified 21 Finnish Web-based auctions in the December 2000 through May 2001 timeframe.

There are two types of business models employed by online auction sites. First, there are the so-called listing-agent sites that act as an agent for other sellers, allowing them to register their items and running the auctions on their behalf (Lucking-Reiley, 2000). The other primary business model for Internet auctions is called a merchant site, where an auction site itself acts as a retailer. This business model, however, is not very common, at least in the Finnish market. Internet-based online auctions typically use the English auction model (or ascending auction); other major auction models are the Dutch auction (or descending auction), first-price and second-price sealed-bid auctions, and the double auction (Kambil & van Heck, 2002).

Process Approach to Online Auctions

Our analysis will be based on the model of exchange processes developed by Kambil and van Heck (1998). The current version of the model was introduced by the authors in 2002 (see Figure 1). The three main elements of the model of exchange processes are the basic trade process (i.e., search, pricing, logistics, payment and settlement, and authentication); the trade context processes (i.e., product representation, regulation, risk management, influence, and dispute resolution); and finally, the communications and computing element facilitating and supporting the two sets of exchange processes.

Figure 1. Exchange process model (Kambil & van Heck 2002)

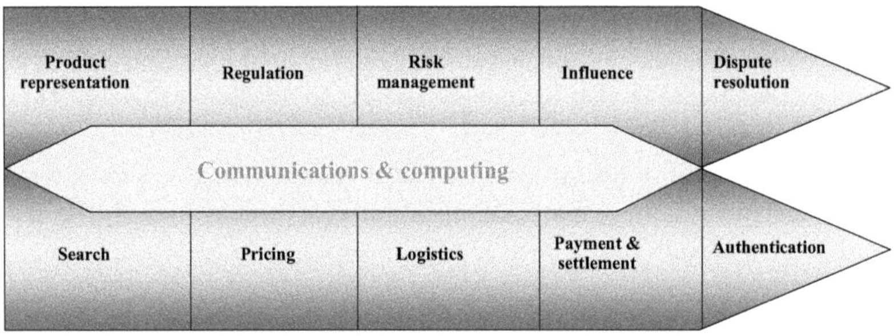

The model of exchange processes has been developed and tested in the context of traditional and online auctions, which makes it a suitable process framework for our analysis of the relationship between multi-channel technologies and exchange processes in Finnish online markets.

We will use the Kambil and van Heck model to map the various multi-channel technologies in order to gain a better understanding of their impact on the processes, both individually and in combination with other technologies. By process, we mean "a time-dependent sequence of events governed by a process framework" (McKenzie, 2000, p. 113). In other words, in our research setting, the process refers to an auction process in the context of online auctions, the event refers to an auction transaction, and the process framework is the exchange process model developed by Kambil and van Heck (1998, 2002).

Conceptual Framework and Propositions

In order to map the various multi-channel technologies to the Kambil and van Heck model, we utilize a conceptual framework based on theories of marketing channels, multi-channels, and communication technologies (see Figure 2). Within this framework, we have identified four factors that drive the adoption of multi-channel technologies. The four factors are (1) perceived appropriateness of a given multi-channel technology, (2) media richness of various multi-channel technologies, (3) support of multiple modes of communication relationships, and (4) the behavioral trend toward increased use of multiple channels. These

Figure 2. Conceptual framework

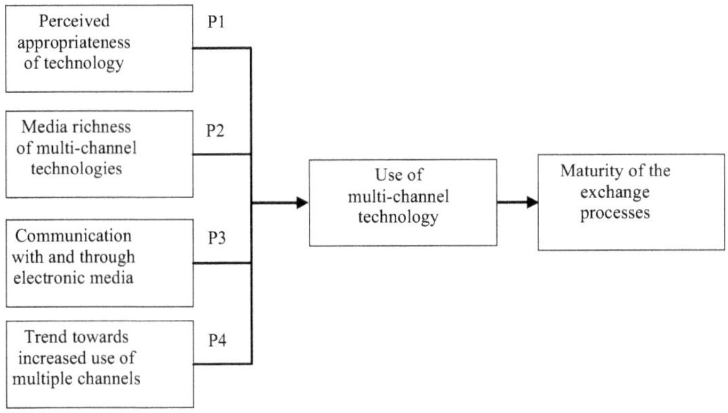

factors will be translated into the four propositions that we will use to analyze the use of multi-channel technologies in the five Finnish online auction sites. The remaining relation in the conceptual framework is the correlation between the use of multi-channel technology and the maturity of exchange processes in online auctions. Earlier research (De Ruiter & van Heck, 2004; Paarlberg, 2001; Vesa & van Heck, 2003) has provided evidence that there is a positive correlation between these two measures.

Perceived Appropriateness of A Given Multi-Channel Technology

According to Griffith and Northcraft (1996), "different implementations and contexts may determine the perceived appropriateness of a given medium" (p. 274). They define media as technology or system, which has a certain objective and psychological characteristic (p. 273). Furthermore, they define media as technology or system, which has certain objective and psychosocial characteristics. Fulk et al. (1990) argued that factors such as media evaluation, task evaluation, and situational factors combine in a complex manner to influence users' perceptions of a communication medium and its use. A positive correlation is expected between the perceived appropriateness of a given multi-channel technology and its use in conjunction with various exchange processes. The observations discussed suggest the following proposition:

> **Proposition 1.** The more appropriate a stakeholder (i.e., buyer, seller, or market maker) perceives a given multi-channel technology for

performing a given exchange process, the more it will use multi-channel technologies offered by online markets.

Media Richness of Various Multi-Channel Technologies

Lengel and Daft (1984) argue that some media are richer than others, if we compare the ability of information provided by a given medium to change understanding within a time interval. Online auctions are using only electronic channels that cannot compete with richer media such as face-to-face contact in a physical auction facility or telephone conversation with an auctioneer. This view is supported by Connolly et al. (1990) who highlighted the "impersonal, anonymous character of electronic communication in general, suggesting that social cues necessarily seem to be decreased by the use of electronic media" (p. 5). This leads us to anticipate that there is a negative correlation between the use of multi-channel technologies and the quality of communication when measured in terms of media richness, which leads us to the following proposition:

Proposition 2. The inability of electronic channels to support rich communication reduces the use of multi-channel technologies in online auctions.

Support Multiple Modes of Communication Relationships

Hoffman and Novak (1996) have identified three different communication relationship types: consumers can interact with the medium (e.g., surf the Web); firms can provide content to the medium (e.g., a firm establishes a Web server); and the consumer can create content in the medium (note that the World Wide Web is defined as a medium in this content). Furthermore, Hoffman and Novak (1996) note that the Internet also can be used for computer-mediated communication among consumers and/or firms through the medium. A positive correlation is expected between the multi-channel technologies' ability to support multiple modes of communications and the use of those technologies.

Proposition 3. The ability of the stakeholders of an online auction to engage in multiple types of communication relationships increases the use of multi-channel technologies.

Behavioral Trend Toward Increased Use of Multiple Channels

There is a clear trend in the market toward increased use of multiple online channels among consumers (Montoya-Weiss et al., 2003; Reda, 2002; Shim et al., 2004). Nunes and Cespedes (2003) claim that customers are "using all the available channels, entering different ones to fulfill their needs at different stages" (p. 99). As online auctions do not operate in isolation from the rest of the society, it is expected that the behavioral patterns adopted in other service situations and usage occasions will positively affect the use of multiple multi-channel technologies in the context of online auctions.

> **Proposition 4.** The trend toward increased use of multiple channels by consumers increases the use of multi-channel technologies when participating in online auction markets.

Table 1 summarizes the four propositions that we have formulated, based on the review of theories of marketing channel and communication technologies.

Finnish Online Auction Markets

For the purposes of our empirical analysis, we will use the five Finnish online auctions that have been used in our previous study (Vesa & van Heck, 2003). This gives us the opportunity to apply a longitudinal approach in our work. We briefly compare the situation of each online auction from the previous review in February 2003 to their situations in July 2004. Our analysis is based on observation of the selected auctions and functional analysis of their services, using a scoring mechanism developed by Paarlberg (2001). The scoring mechanism was developed further by Vesa and van Heck (2003) in order to gain a better understanding of various technologies used by buyers and sellers to access

Table 1. Summary of propositions

Proposition	Effect	Description
P1	+	Perceived appropriateness of technology
P2	-	Media richness of multi-channel technologies
P3	+	Support multiple modes of communication
P4	+	Behavioral trend toward increased use of multiple channels

the online auction service. The five Finnish consumer auction markets studied are Huuto.Net, Keltainen Pörssi, QXL Finland, Systeemi.Net, and Tori.

- **Huuto.Net** was started in January 1999 by a young entrepreneur, who reserved the domain name and started to run the service on a hosting platform in Florida (Lohikoski, 2002). At the time, eBay had already become very successful in the US, which offered a good reference point for the new service. As Huuto.Net grew very fast, the founder decided to sell it to Bukowski, one of the oldest and most famous auction houses, in August 1999. However, as the Internet boom faded, Bukowski sold Huuto.Net to Sonera, the largest telecom and Internet operator in Finland. Today, Huuto.Net is part of the leading Internet portal called Sonera Plaza. In September 2003, Huuto.Net had over half a million visitors per month. The business model of Huuto.Net is built around advertising revenue, although there is a limited number of services that customers have to pay for (e.g., paid authentication and extra visibility). According to a representative of Sonera, "advertising alone cannot support service like this" (Lohikoski, 2002, p. 10).

  In our earlier study conducted in February 2003, Huuto.Net was clearly the leading online auction in Finland with more than 80,000 items on sale. It was the only online market offering a mobile (WAP) version of their auction service. In July 2004, Huuto.Net is still going strong; the number of items on sale (more than 110,000) is about 100 times more than in the next biggest online auction in this group of companies. Huuto.Net is still the only online market offering a mobile (WAP) version of its services.
- **Keltainen Pörssi** Web auction is a sideshow of the leading printed and online free-ad publication in Finland. The role of this online auction is to support the catalog business. Since our previous audit of this online auction site in February 2003, Keltainen Pörssi Internet auction has remained apparently the same, if we look at the volume of business or the functionality offered. In July 2004, the number of items on sale was well below 1,000 items.
- **QXL** is one of the few international online auctions that operated in Finland during our previous analysis of the five case auctions. The British online auction entered the Finnish market in the summer of 2000 when it bought a Web auction called Bidlet.fi that, at the time, had more than 55,000 customers. During our previous study in February 2003, QXL was operating actively in Finland. However, in July 2004, the Finnish Web site of QXL was temporarily out of use.

- **Systeemi.Net** is a free Web auction mainly for stamps and other collectors' items. In February 2003, Systeemi.Net placed second in Google's page-ranking list for online auctions, albeit the volumes were small. In July 2004, Systeemi.Net is still functioning, with no apparent changes in the scale or functionality of the business.
- **Tori** is a small free online auction site. In February 2003, this auction site held the fifth position in Google's page-ranking list, but on the review date, Tori Web auction did not have a single item on sale. In July 2004, Web auction Tori was still functioning—and still, there were no items on sale.

Lessons Learned

Our findings indicate that Finnish online auctions have built their businesses on a single channel—the Internet. Alternate technologies, such as mobile data services, digital TV networks, and set-top boxes, have not been implemented, although the mapping of various multi-channel technologies and exchange processes suggests, for instance, that mobile services could offer added value for various stakeholders. The fact that Finnish online auctions support only the Internet channel is surprising, if we look at the penetration rate of various multi-channel technologies in Finland. While approximately 40% of Finnish people between 15-79 years of age used the Internet daily or almost daily in the period of January-February 2004 (source: Taloustutkimus Oy), the mobile phone penetration rate in Finland at year-end 2003 was already 94% of households (source: Tilastokeskus). Based on these figures, one would expect to see more implementations of mobile technology in the context of online auctions. We believe that our study will help service providers to identify the processes in which mobile technology would offer the biggest benefits. In this section, we go through the four research propositions developed in Section 3 in order to see if they are supported by our empirical findings.

> **Proposition 1.** Increased perceived appropriateness of a given multi-channel technology leads to increased usage.

The first proposition stated that the more appropriate a buyer or seller perceives a given multi-channel technology for performing a given task, the more it will be used (i.e., a positive impact). Different technologies have different characteristics, making them more or less suitable for performing various exchange processes. Based on the characteristics of the multi-channel technologies reviewed here, one would expect, for instance, that digital television would be a superior technology for performing the product representation process, as digital television offers a rich combination of text, images, audio, and video (Hoffman & Novak 1996), not to mention the fact that, for most people, using the TV would offer a much more natural context of use than sitting in front of a PC while browsing various items on sale (albeit, in this chapter, we do not focus on the context of use while participating in online auctions). However, as a digital television system in Finland does not yet offer a return channel (i.e., in Finland digital television is still mainly about broadcasting), interactive services are too complicated to implement from market makers' points of view. Furthermore, mobile phones equipped with appropriate data services would support well the intense bidding phase of an online auction, allowing buyers to participate in online auctions independent of time and place. However, although the penetration of mobile phones is among the highest in the world, so far only one online auction in the whole country offers this option.

Despite the obvious strengths of different multi-channel technologies, our analysis shows that the buyers and sellers do not have the possibility of choosing the most appropriate technology in the Finnish online auction, because most online auctions support only the Internet channel. Even in the only auction that offered mobile access to exchange processes, the functionality of the WAP version of the service was very limited.

One could argue that the market maker has made this decision on customers' behalf, most likely based on financial calculations and less likely on what customers might find most suitable. Also, the size of the Finnish market (i.e., a population of a little more than 5,000,000 people) may have an impact on this, as the cost of implementing, for instance, a mobile version of a Web-based auction site creates additional costs. The low penetration rate of interactive set-top boxes probably explains why market makers have not introduced online auction services in digital television. However, in the fall of 2004, the Finnish consumer electronics company Finlux announced their new digital broadband TV, which combines digital television and broadband Internet access in the same device. The move is justified by the assumption that digital television soon will become much more interactive, which also could lead to an increased use of digital TV as a service delivery channel for online auctions.

Based on the previous discussion, we are forced to conclude that the first proposition is not supported, although the real reason behind the current situation is more likely the lack of alternatives rather than the lack of interest in using a combination of multiple electronic channels: It is difficult to choose the most appropriate technology for a given task (or, in this case, for a given exchange process) when there are no options to do this. This fact is often ignored in multi-channel research, as one of the assumptions behind the research setting is that the users have the possibility to choose between different channels—or, perhaps even more often, they are asked to predict their behavior, if there were multiple electronic channels available for performing a given task.

Proposition 2. Lower media richness reduces the use of electronic channels.

The second proposition argues that the inability of electronic channels to support rich communication reduces the use of multi-channel technologies. The limitations of electronic forms of communication have been pointed out in several earlier studies (Connolly et al., 1990; Griffith & Northcraft, 1994; Lengel & Daft, 1984). The findings suggest that communication over electronic media has less social cues, is less immediate psychologically, and is not as efficient in problem solving as non-electronic communication.

Our analysis indicates that electronic multi-channel technologies had a very limited role in supporting those exchange processes that had to deal with more complicated issues such as dispute resolution. On the other hand, when dealing with numeric information, such as price of products or a new bid, multi-channel technologies are performing very well (i.e., in the context of the basic trade processes as described by Kambil and van Heck, 2002). This finding is not surprising, because earlier research has demonstrated that in situations where the level of uncertainty (absence of information) and the level of equivocality (multiple and conflicting interpretations of the situation) are high, richer media are often needed. There are, however, certain areas where new technology enables new forms of more collective ways of communications. For example, an online auction's reputation system supported by online chat room functionality gives the buyers and sellers an opportunity to create social pressure on misbehaving participants. In traditional auctions, it would have been difficult to offer such an efficient feedback system via a traditional word-of-mouth approach.

Once again, we can only draw conclusions regarding the Internet channel, as other multi-channel technologies have not been implemented in Finnish online auctions yet. Based on the functional analysis of the exchange processes of these selected online auctions, we conclude that our second proposition is supported by the empirical findings; in other words, the limitations of communication richness in electronic channels reduces the use of the technologies in the context of more complex trade context processes (Kambil & van Heck, 2002).

Proposition 3. Multiple communications relationships increase the use.

The third proposition stated that the ability of stakeholders to engage in multiple types of communication relationships increases the use of multi-channel technologies (i.e., positive correlation). Our analysis shows that an online auction offers a rich communication environment for sellers, buyers, and the market maker, who all have the opportunity to communicate both with the medium (i.e., the auction Web site) and through the medium (e.g., e-mails or SMS to each other, or participating in an online chat on the auction site). Innovative implementations of multi-channel technology, such as a reputation system on the Internet or online chat, can offer rich communication, even though the media are electronic.

Although one might easily think that an online auction site is yet another Web-based service for consumers to surf through every now and then, it is important to bear in mind that in markets—online or off-line—the role of the people involved in auction transactions is essential. According to Kambil and van Heck (2002), "human beings have always made markets, and they will continue to do so in the future—even if business is conducted through cyberspace" (p. 3). Therefore, it is justified to say that, at the end of the day, online auctions are about communication between buyers and sellers, and within the online auction community, in general, the role of communications technology is to support this: "Electronic markets are not technological interactions supported by humans. They are human interactions supported by technology" (Kambil & van Heck, 2002, p. 3).

If we also keep in mind that short-message service (SMS) has been, by far, the most successful non-voice mobile service in Finland, and that e-mail has been the driving force behind the growth of PC use and Internet connections in households, it is easy to understand why multi-channel technologies often are used for communication also in the context of online auctions: Asking for additional information through the auction Web site (one-to-many) or directly from the seller by sending an e-mail or SMS (one-to-one); making arrangements for the delivery and payment of the product by sending an e-mail or by calling (one-to-one electronic or non-electronic communication); or checking other buyers' experiences of a seller in an online chat or bulletin board (many-to-many computer-mediated communication). These are only few examples of the numerous ways in which multi-channel technologies are used when various stakeholders interact with the auction site and with each other. This discussion leads us to the conclusion that proposition 3 is supported by our empirical findings.

> **Proposition 4.** A trend toward increased use of multiple channels in other industries leads to increase use of multi-channel technologies in auctions.

Proposition four argues that the current trend (i.e., in retail and banking) of increased use of multiple channels also leads to an increased use of multi-channel technologies in online auctions. Finnish consumers use their mobile phones very actively for ordering ringtones and screen savers, not to mention all the SMS messages sent almost daily. On the other hand, Finnish banks have been very successful in moving bill payments and other financial transactions to electronic channels. Against this background, one would expect that Finnish consumers have experienced the current trend toward a multi-channel philosophy when dealing with other types of services.

Once again, our empirical data works against our intuition. Although we believe that a positive correlation exists between the general trend toward a multi-channel approach in other industries and the use of multi-channel technologies in online auctions, as suggested, for instance, by innovation diffusion theories (Rogers, 1995), we could not find support for this proposition. The reason is that the Finnish online auctions typically do not offer more than one channel; thus, the trend toward the use of multiple channels cannot be verified in the context of Finnish online auctions. Therefore, we have to conclude that proposition 4 is not supported.

Our analysis shows that Finnish online auctions, to a great extent, are applying a single-channel strategy by supporting only the Internet channel. Despite the high penetration rate of mobile phones in Finland, the mobile channel is used only by one online auction house. One explanation for this could be that the current mobile handsets and networks are suffering from poor user-interface (i.e., small screen size) and slow data transfer rates, respectively. The results also show that the trend predicted by consumer behavior researchers toward increasingly multi-channel oriented consumers (Shim et al., 2004), who actively surf between various channels (Nunes & Cespedes, 2003), does not seem to be the case in this context.

In one area, the Finnish online auctions are doing reasonably well. The buyers and sellers communicate actively with the auction site but also with each other by using multi-channel technologies—or, at least, multiple applications within one technology. For instance, the Internet is used to access the World Wide Web with a browser, to send e-mail messages during the bidding process and after the deal is closed, and to chat online to share experiences with other users. Likewise, mobile phones are used during the bidding process, but also to establish a contact between the buyer and the seller, once a deal has been made either by sending SMS messages or by calling. One could argue that this phenomenon is very typical in Finland, where person-to-person communication, for instance, by SMS has been a huge success, while person-to-content services typically have failed. This study raises the question why the Finnish online auctions are performing so poorly compared to the leading international players. There are several potential explanations for this situation:

- Finnish consumer protection and remote sales legislation prevents auction houses from getting involved in the auction process, and, therefore, they are not in the position to develop the exchange processes.
- Developing new multi-channel services is expensive, and the Finnish markets are too small for these kinds of investments.
- Finnish people do not take auctions very seriously, and there might be a cultural barrier in that sense.
- There is a lack of dominant players in the Finnish auction industry, as eBay is presently not present in Finland (due to regulatory issues).

We believe that testing our conceptual model and the set of propositions developed here in more advanced markets would provide more results.

Conclusion and Further Research

This chapter provides two key contributions to the literature of multi-channel technologies and online markets. The first contribution is that this chapter develops a new conceptual model and a set of propositions for the adoption of multi-channel technologies, based on the theories of marketing channels, multi-channels, and communication technologies. These important factors are perceived appropriateness of a given multi-channel technology, media richness of various multi-channel technologies, support of multiple modes of communication relationships, and the behavioral trend toward increased use of multiple channels.

The second contribution is that these factors were analyzed in five online consumer auction markets in Finland. The results of the analysis show that media richness of various multi-channel technologies and the support of multiple modes of communication relationships are important in the adoption and use of multi-channel technologies in the Finnish auction markets. The analysis also shows that these auction markets still are not very advanced in using sophisticated multi-channel technologies.

Further research will compare the adoption of multi-channel technologies by Finnish consumer auction markets with Dutch consumer auction markets (De Ruiter & van Heck, 2004). The next step will be to test the specified propositions. This could be realized by performing a large-scale research on European online consumer auction markets. As a result, it could be possible to make statements, based on in-depth quantitative results, about whether the proposed hypotheses are valid or not. Furthermore, these statements could be generalized over all online consumer auction markets.

References

Bakos, J.Y. (1991). A strategic analysis of electronic marketplaces. *MIS Quarterly, 15*(3), 295-310.

Bakos, Y. (1998). The emerging role of electronic marketplaces on the Internet. *Communications of the ACM, 41*(8), 35-42.

Balderston, F.E. (1958). Communications networks in intermediate markets. *Management Science, January, 4*, 154-171.

Bapna, R., Goes, P., & Gupta, A. (2003). Analysis and design of business-to-consumer online auctions. *Management Science, 49*(1), 85-101.

Clark, T.H., & Stoddard, D.B. (1996). Interorganizational business process redesign: Merging technological and process innovation. *Journals of Management Information Systems,13*(2), 9-28.

Connolly T., Jessup, M., & Valacich, J.S. (1990). Effects of anonymity and evaluative tone on idea generation in computer-mediated groups. *Management Science, 36*,97-120.

Daft, L.R., & Lengel, R.H. (1986). Organizational information requirements, media richness and structural design. *Management Science, 32*(5), 554-571.

De Ruiter, S., & van Heck, E. (2004). The impact of multi-access technologies on consumer electronic auctions. *European Management Journal, 22*(14), 377-392.

Etgar, M., & Zusman, P. (1982). The marketing intermediary as an information seller: A new approach. *Journal of Business, 55*(4), 505-515.

Griffith, T.L., & Northcraft, G.B. (1994). Distinguishing between the forest and the trees: Media, features, and methodology in electronic communication research. *Organization Science, 5*(2), 272-285.

Hoffman, D.L., & Novak, T.P. (1996). Marketing in hypermedia computer-mediated environments: Conceptual foundations. *Journal of Marketing, 60*, 50-68.

Kambil, A., & van Heck, E. (1998). Reengineering the Dutch flower auctions: A framework for analysing exchange organizations. *Information Systems Research, 9*(1), 1-19.

Kambil, A., & van Heck, E. (2002). *Making markets: How firms can design and profit from online auctions and exchanges*. Boston, MA: Harvard Business School Press.

Laine, J. (2002). *The rules of eAuctions* (in Finnish). Research report, LTT Research Ltd., Electronic Commerce Institute. Helsinki, Finland.

Lohikoski, L. (2002). *Huuto.Net—Business and technology*. Presentation at Helsinki University of Technology, Espoo, Finland.

Lucking-Reiley, D. (2000). Auctions on the Internet: What's going to be auctioned and how? *Journal of Industrial Economics, 48*(3), 227-252.

Malone, T.W. (1995). Inventing the organizations of the 21st century: Control, empowerment, and information technology. *Proceedings of the Harvard Business School Colloquium, Multimedia and the Boundary-Less World,* Boston, November 16-17.

Malone, T.W., Yates, J., & Benjamic, R.I. (1987). Electronic markets and electronic hierarchies. *Communications of the ACM, 30*(6), 484-497.

McDonald, C.G., & Slawson Jr., V.C. (2002). Reputation in an Internet auction market. *Economic Inquiry, 40*(3), 633-650.

McKenzie, K.D. (2000). Processes and their frameworks. *Management Science, 46*(1), 110-125.

Montoya-Weiss, M., Voss, G.B., & Grewal, D. (2003). Determinants of online channel use and overall satisfaction with a relational, multichannel service provider. *Journal of the Academy of Marketing Science, 31*(4), 448-458.

Nunes, P.F., & Cespedes, F.V. (2003). The customer has escaped. *Harvard Business Review, 81*(11), 96-105.

Paarlberg, J.A. (2001). *Web auctions in Europe: An analysis of 194 consumer Web auctions in eight European countries* [master's thesis]. Rotterdam: Erasmus University of Rotterdam.

Puhakainen, J., & Tuunainen, V. (2001). *Consumer-oriented e-auctions in Finland* [in Finnish]. Research report, LTT Research Ltd., Electronic Commerce Institute, Helsinki, Finland.

Reda, S. (2002). Active multi-channel shoppers may be a liability, less loyal that other on-line shoppers. *Stores, 84*(9), 78.

Reve, T., & Stern, L.W. (1979). Interorganizational relations in marketing channels. *Academy of Management Review, 4*(3) 405-416.

Rogers, E.M. (1995). *Diffusion of innovations*. New York: The Free Press.

Roth, A.V., & Jackson, W.E. (1995). Strategic determinants of service quality and performance: Evidence from the banking industry. *Management Science, 41*, 1720-1733.

Shim, S., Eastlick, M., & Lotz, S. (2004). Search-purchase (S-P) strategies of multi-channel consumers: A segmentation scheme. *Journal of Marketing Channels, 11*(2/3), 33-54.

Steuer, J. (1992). Defining virtual reality: Dimensions determining telepresence. *Journal of Communication, 42*(4), 73-93.

Strader, T.J., & Ramaswami, S.N. (2002). The value of seller trustworthiness in C2C online markets. *Communications of the ACM, 45*(12), 45-49.

Trevino, L.K., Daft, R.L., & Lengel, R.H. (1990). Understanding managers' media choices: A symbolic interactionist perspective. In J. Fulk, & C. Steinfield (Eds.), *Organizations and communication technology* (pp. 71-94). Newbury Park, CA: Sage.

Vesa, J., & van Heck, E. (2003). The impact of multi-access technologies on exchange processes in five online auction markets in Finland. *Proceedings of the Post-Graduate Seminar,* Turku Centre for Computer Science, TUCS General Publication, No. 25, June, Turku, Finland.

Section III

Envisioning the Future of the Digital World

Chapter XVIII

The Changing Role of Middle-Men Infomediaries

Mirella Lähteenmäki, Helsinki School of Economics, Finland

Markku Tinnilä, Helsinki School of Economics, Finland

Abstract

Intermediaries and especially information-intensive infomediaries have been hailed as winners of the new electronic economy. The need for infomediaries arises from the reorganization of several markets as a result of the Internet. The main claim of emerging e-business companies during the hype at the turn of the millennium was shortening of the value chains. This meant bypassing middle-men with a direct contact to end-customers, in many cases, consumers. However, the millions of Web sites failed to attract customers, simply because they could not be found among all the others. The solution to this problem was the reintroduction of middle-men actors; in this case, a new type of player, the infomediary, focusing on gathering, organizing, and utilizing essential customer information. The infomediary maximizes the value of information by helping clients to locate relevant products and services based on their individual needs and wants or on their personal profiles. In this chapter, we intend to provide a typology of infomediaries, their roles, and value creation models, and demonstrate infomediary evolution by analyzing exemplary cases.

Intermediation

The intermediaries' basic role is to improve the efficiency with which products and services move through channels to customers by reducing the transaction costs associated with bringing a product to market. Every market-related function can be handled by an exterior channel member. However, emerging network technology has and will alter the role of intermediaries, as traditional companies (i.e., computer businesses, etc.) struggle with channel conflicts and disintermediation (Doyle, 2000) when customers find it ever easier to bypass traditional mediaries by directly contacting end customers. On the other hand, in many information-intensive fields, technology encourages intermediation, and new kinds of channel members, information intermediaries, or infomediaries emerge rapidly.

But what, then, is an infomediary? The new term *infomediary* originally was introduced by John Hagel and Jeffrey Rayport in their 1997 article entitled, "The Coming Battle for Customer Information" in *Harvard Business Review*. In the consumer marketplace, infomediaries are simply agents that facilitate Internet commerce by transmitting and manipulating personal customer information, thus competing with their ability to capture, utilize, and manage data in a manner that adds value for clients, buyers, or sellers. These virtual actors appear in many forms, ranging from plain platforms, where customers can exchange goods with each other (e.g., virtual auction sites), to versatile service providers offering complex service packages, varying from custom search agents to personal customer information management and consultancy services.

Despite growing intermediation or infomediation, literature on mediaries still lacks solid conceptual and empirical research; most articles and writings have been prepared rather as a basis for discussion (Moon, 2000). In addition, proposed infomediary categories quite often are overlapping or otherwise confusing. We are aiming to partly fill that gap by providing a typology of infomediaries, their roles, and value creation models, and then demonstrating infomediary evolution by analyzing exemplary cases.

Infomediary Dimensions

Infomediaries are best characterized by their customers and service offering, ranging from consumers (client-centric infomediary) to marketers (vendor-centric infomediary) (dimension 1), and from information platforms to complex services packages (dimension 2), as shown in Figure 1. Next, our purpose is to analyze these dimensions more specifically first by studying marketer platforms

Figure 1. Infomediary dimensions

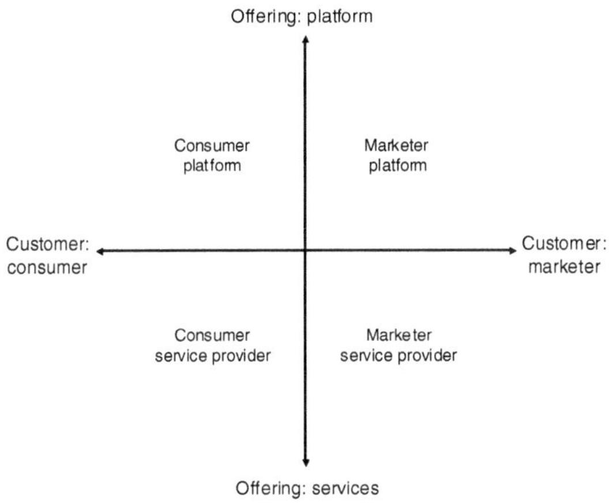

and services (right side) and then consumer platforms and services (left side). Infomediaries also can take a neutral position and serve both buyers and sellers by matching their needs (i.e., generic agents). However, quite often, matching services start to focus either on buyers or sellers and, thus, lose their neutral position.

Marketer Platforms and Services: Seller Agents

Most infomediaries follow business models that generate revenue from the seller side (Grover & Teng, 2001). Relationships between these two parties can be open (non-proprietary relationship) or closed (proprietary relationship) (Grover & Teng, 2001). In the case of an open relationship, anyone willing to pay has access to services provided, whereas in the case of a proprietary relationship, only certain named members have access to services. In both cases, the common main task is to provide added value and help marketers bring products to market, where customers insist on value and relevant benefits instead of products.

Customized services are becoming increasingly important to customers. Van Raaij and Poiesz (2003), for example, predict that, in the future, products and services will no longer be marketed on a stand-alone basis with no additional benefits attached. Rather, products and services of different domains will be combined into functional all-in-one packages. For example, marketers don't sell apartments anymore but rather a possibility to live comfortably, including interior decoration, garden architecture, maintenance, house repairs, cleaning, safety, and monitoring during owners' absences. The package provider, the infomediary, eventually becomes a trusted brand for a group of loyal customers adapting the product package to their needs and wants. Archol and Kotler (1999), for example, refer to brand mediation. The service provider strives for relationships that cover customers' lifetimes; customers are followed over life stages, and products and services are carefully attuned to these different stages.

Services provided by marketer information mediaries vary quite substantially. On the other side of the continuum, marketers offer plain platforms (upper right-hand corner in Figure 1), where retailers can cheaply sell their products and services or just provide product information. For example, most auction sites simply enable transactions and charge a percentage fee for each product sold. However, marketer service providers also offer more complex services (lower right-hand corner in Figure 1). For example, directories (industry magnet Web sites) combine smaller players and complementary products and services by screening suppliers, evaluating and comparing products and services, and facilitating transactions (Grover & Teng, 2001).

Consumer Platforms and Services: Buyer Agents

Just a few infomediaries follow business models that generate revenue from the buyer side. The reason is simple: customers are reluctant to pay for Internet services, especially when providers are small and unknown. Still, consumer information intermediaries exist and can be expected to increase over the next few years, based on the steadily growing need and value for personal customer information. Before, traditional direct marketers were among the few businesses that tracked individual customers and purchases using computer databases. Today, advances in information technology, fragmented media audiences, and demands for economic efficiency have spurred a much broader use of consumer information (Phelps et al., 2000), and growing numbers of marketers routinely collect and utilize personal customer information for relationship marketing purposes.

In order to obtain personal customer data, marketers have to persuade customers to share their attitudes, preferences, and purchase patterns with them on a regular basis (Wind & Rangaswamy, 2001). However, several factors, especially privacy concerns, have resulted in growing consumer reluctance to provide any personal data. Also, the value offered for personal customer information is questionable; consumer information is a valuable business asset, and yet, its owners (consumers) receive nothing for it in the current marketing model (Archol & Kotler, 1999). According to several authors (Fournier et al., 1998), consumers should feel that they really benefit by providing personal information to marketers. Infomediaries, controlled by customers and relying on customer ownership of information, might solve the problem.

Consumer information mediaries are expected to appear in many forms. In the upper left-hand corner of Figure 1, consumer platforms are simple market spaces or chat areas, where consumers share information, opinions, and attitudes on products, services, and their use. In other words, infomediaries enable clients to create content by providing them an effective platform. However, most consumer platforms gradually have evolved into more complex service providers, offering themselves content and not just a platform for exchange (lower left-hand corner of Figure 1). For example, infomediaries can provide customers with intelligent agents that retrieve accurate and relevant information from the Web, compare possible options, and rate them trustworthy.

After some time and intensive interaction and exchange, infomediaries (i.e., search agents) know consumers thoroughly and can propose suitable products and services. Ultimately, the mediary becomes a personal assistant with a profound knowledge of his or her boss—the consumer. The agents assist consumers with relevant product information and alerts about deadlines. Areas of possible assistance include information management and travel and housing assistance. Agents can monitor the situation of the customer and find interesting and affordable improvements. Agents also can learn from the feedback they get from customers to improve their services (Van Raiij & Poiesz, 2003).

Typology Creation

Infomediaries have existed in various forms, including search engines, communities of special interest, industry magnet sites, e-tailers, and even corporate Web sites (Grover & Teng, 2001). However, the extent to which they provide mediary functions varies, as do the various business models on which they tend to operate. However, based on our grouping (customer vs. offering), infomediaries can be categorized into marketer and consumer platforms and marketer and consumer service providers. The classification is overlapping, but it is a suitable description for the current situation.

Marketer Platform

Marketer platforms offer their clients or marketers a virtual market place or market space, thus radically reducing obstacles of profiting from consumers' niche needs. Aggregating a large number of sellers on a common Web site, vendor-centric marketer platforms, or market makers, enable marketers to identify people with highly specialized needs and communicate with them efficiently. Sufficient traffic is created by diminishing consumers' search costs and saving their energy and time. Practically, marketer platforms have to offer functions such as search engines, directories, and comparison (Grover & Teng, 2001). A good example of a marketer platform is eBay, the popular auction Web site and pioneer of online trading.

Consumer Platform

When consumers themselves create content, marketer platforms turn into consumer platforms—virtual communities. Their principal objective is to assist clients or consumers with similar buying needs by enabling and creating discussion (i.e., chat areas, etc.) on products, services, and vendors, thus helping them to compare offerings, choose among sellers, and switch from one vendor to another. Marketers can participate in consumer platforms by organizing and managing communities, thus gaining revenue from advertising, transactions, and subscription (Hagel, 1999). This marketing approach is called *collaboration marketing* and enables marketers to learn more about their customers. Customer platforms also have relevance for business-to-business marketing evolving typically around an industry (e.g., Physicians' Online).

Marketer Service Provider

Another category of infomediaries consists of players that assist other companies in creating, enhancing, and sustaining customer relationships by handling and utilizing personal customer information. Traditionally, information handling providers offering services to marketers (vendor-centric) fall into two basic types (Hagel & Rayport, 1997): audience brokers (e.g., Doubleclick), who use their knowledge of Web-page audience composition to help advertisers reach the right audiences; and lead generators (e.g., Autobytel.com), who aggregate potential customers according to their properties, behavior, and preferences, and then direct customers to vendors whose offerings meet their needs. For example, Engage Technologies places cookies on personal computers in order to track consumer behavior over time and then creates profiles of these users based on their browsing patterns and personal interests (Moon, 2000). Personal profiles are sold later to marketers. A more sophisticated case is Nectar, a unique reward program that lets consumers collect points at various places by using a single card. Although established by Sainsbury's (daily consumer goods), Debenhams (department store chain), Barclays (bank), and British Petroleum (energy company), Nectar is an independent company managing all personal customer information associated with the card. Revenue is gained by selling information to the members of the network (e.g., Sainsbury's).

Consumer Service Provider

Consumer service providers specialize in collecting and utilizing personal customer information for relationship marketing purposes and negotiating, coordinating, and brokering transactions for the clients' benefit (client-centric), thus remaining independent of suppliers. Infomediaries create customer value (Archol & Kotler, 1999; Grover & Teng, 2001; Hagel & Rayport, 1997) by (1) finding relevant marketers and offerings, especially with complex features (agent function); (2) representing consumers in negotiations with vendors seeking personal customer information (proxy function); and (3) screening commercial messages for customers (filter function). In addition, standard regulation as well as privacy and security services are provided. Possible revenue sources include payments made by customers for services rendered (i.e., annual fee, etc.) or a commission on revenues accruing to customers from marketers as payments for personal customer information (cash, etc.).

Cases

Next, we will present and analyze some exemplary cases based on the previous grouping in marketer platforms and service providers and consumer platforms and service providers.

Case 1 (Marketer Platform): eBay

Founded originally in September 1995, eBay pioneered online trading by developing a virtual market space, or marketer platform, where buyers and sellers, both individuals and businesses, can get in touch with each other in an entertaining auction format. With more than 27,000 product and service categories, it is one of the most popular shopping Web sites. eBay is an excellent example of a marketer platform; it provides customers (sellers) an efficient trading platform by mediating information and knowledge on products and services, and it generates revenue from fees ($0.30 to $3.30) and commissions (1.25% to 5%) on items sold.

eBay's success has proven that information mediation is a possible business source. eBay acts as a transaction enabler by bringing buyers and sellers together without actually touching the items sold. Avoidance of logistical problems probably has been the key to eBay's incredible success. However, as Hagel and Rayport (1997) predicted, eBay has gradually diversified its offering from its basic platform by starting to provide various selling (e.g., Paypal protection program), bidding (e.g., retraction possibility), and buying (e.g., warranty program) services, depicted as vertical movement in Figure 2.

eBay's future challenges include serving giants (Sears, etc.) without angering the original smaller entrepreneurs, making selling easier (listing, etc.), marketing and appealing to the big crowds and masses, and especially enhancing consumer trust. In consumer Internet marketing, especially virtual auctions, the role of trust is extremely important, as reports of frauds have been rising. One possible solution is discussion groups and boards, where customers meet with each other, discuss topics of mutual interest, and provide one another with information on buying and selling as well as on buyers and sellers (rating system), resembling a consumer platform, thus causing horizontal movement (Figure 2).

Case 2 (Consumer Platform): Baby Magazine Chat

Baby Magazine Chat is an active consumer community, or consumer platform, run by *Sanoma Magazines Finland*, and is based on *Baby Magazine* (Vauva-lehti). It is formed around consumers interested in children and baby care. Most members are mothers of newborn children, who are desperate for information and knowledge. However, most mothers stay members also after their children have grown older, and they create new groups (e.g., teenagers). Also, fathers of children can participate. Baby Magazine Chat is a typical increasing-returns business, becoming more valuable as more and more customers get involved; the value is in the number of people participating.

From the marketer's (*Sanoma Magazines Finland*) point of view, Baby Magazine Chat helps them to understand their customers and learn more about them (collaboration marketing): Mothers are involved in creating content, because baby care is such a significant part of their lives. Some revenue also is gained from advertising and transactions (horizontal movement in Figure 2), though a Web site (even popular sites) is a difficult medium. However, the most important goal is to sustain and deepen relationships with old customers and attract new clients.

Case 3 (Marketer Service Provider): Comma Finland

Established in 1993, Comma Finland is a private organization specializing in travel publishing (magazines) and communications. Its publication portfolio includes

LomaSuomi (*Holiday Finland*) magazine, one of the largest travel media in Finland and other Nordic countries. *Holiday Finland* magazine is published twice a year with a print run of 800,000 copies. Another Comma Finland magazine, *Look at Finland*, is targeted at international audiences and customers. It is published in five languages and currently has a print run of 300,000 copies. Additionally, Comma Finland through the years has produced hundreds of publications, including Finnish Tourist Board brochures.

Comma Finland also is a market leader in online travel communications in Finland. The Finnish-language www.lomasuomi.fi (*Holiday Finland*) travel portal features more than 15,000 travel facts about Finland. Consumers also can get information on their mobile phones. For corporate customers, Comma Finland provides an online photo office (www.comma.fi) with more than 20,000 pictures of Finland. The most recent service for business customers is Finland.com travel portal.

Comma Finland is an interesting example of a marketer service provider (Figure 2), assisting business customers (e.g., hotel chains, tourist areas, etc.) in creating, enhancing, and sustaining customer relationships, both online and offline. Following the marketer service provider typology (Hagel & Rayport, 1997), Comma Finland can be seen as a Web site audience broker (www.lomasuomi.fi) with deep knowledge (based on compulsory registration) of Web page audience composition and with efficient channels (business travel portal) to use that knowledge for the business customers' benefit. With its consumer magazines and travel portals, Comma Finland operates also as a lead generator, directing readers to vendors.

Case 4 (Consumer Service Provider): Helsinki Virtual Village

Helsinki Virtual Village (HVV) portal provides public services (i.e., forms, etc.) and information (i.e., directories, timetables, magazines, etc.) for interested Web site visitors. It is a locally focused portal, specializing in the new Arabianranta area and, thus, serving mainly local residents, students, workers, and companies. For example, a calendar provides information on current local events. However, the primary target group is residents that are also actively participating in the development process of their own portal.

One of the most interesting functions of the HVV portal is housing company Web sites featuring residents' participation and ownership. All residents are offered the membership of their own housing company Web site or group. Memberships as well as Web sites are controlled by an appointed moderator, or virtual janitor, who is mostly a resident. The Web sites include internal information (produced

Figure 2. Infomediary future development

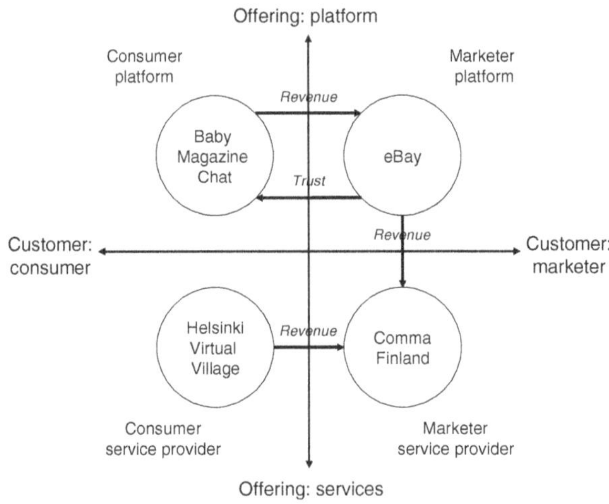

by residents) ranging from board meetings to reservations of common premises. At the moment, residents are planning joint purchases aimed at better prices and easing trouble, which makes the HVV portal an excellent example of a consumer service provider infomediary.

Residents and other consumers' needs and interests are assessed by compulsory registration. Most residents have given personal data willingly, knowing that the data will be used to their own benefit. For example, the city can develop local services based on their needs. Also, companies and service providers get important information on customer needs and wants. In fact, HVV offers several services to local companies and services in order to generate more revenue depicted as a horizontal movement (Figure 2). For example, companies can outsource their Web sites. In this way, linked to the portal, smaller companies especially may get enough attention that would otherwise be impossible.

Conclusion

We have developed and described a categorization for depicting, analyzing and understanding infomediaries. Based on exemplary case companies shown in Figure 2, infomediaries' roles and tasks are developing constantly as they seek new opportunities and sources of revenue. Consumer platforms (e.g., Baby Magazine Chat) and service providers (e.g., Helsinki Virtual Village) especially tend to move toward the direction of marketer platforms and marketer service providers, as consumers are reluctant to pay (enough) for Internet services. However, this is a common problem for all consumer market actors, whether they are involved in retail of goods, services, or information. The only possible solution is to provide relevant services and acceptable pricing methods, as well as to promote customer accustomedness to new Internet services. Clearly, the hectic search for new infomediary business models has not yet been very successful. However, there are signs of positive yet moderate development.

Based on exemplary case companies, trust creation also seems to be a puzzling element, especially for marketers. Yet, for infomediaries, it is crucially important when attracting new customers and developing relationships with them. Researchers, such as O'Malley, Patterson, and Evans (1997) view trust as a compulsory relational element. Looking at customer information-intensive infomediaries, one is left to wonder whether trust has been given the weight it deserves when designing new services, especially for consumers. One way to gain consumer trust is to encourage customers to share their opinions with other customers, a party they certainly trust. For example, Baby Magazine Chat uses this strategy, as well as eBay (Figure 2). Given this fact, it is certainly not surprising that many marketplaces promote customer communities and try to turn community members into customers.

To reiterate the main task of an infomediary, the infomediary maximizes the value of information by helping clients to locate relevant products and services based on their individual needs, wants, or personal profiles. Based on exemplary case companies, this task is both complicated and demanding, requiring a large range of services, an efficient service provider network, trust, a possibility to create and utilize profiles, and so forth. Some might actually declare it a mission impossible. However, while no single infomediary can completely meet these challenges yet, there are quite promising services that are at least partly trying to fulfill the vision prophesied for infomediaries.

References

Archol, R.S., & Kotler, P. (1999). Marketing in the network economy. *Journal of Marketing, 63*, 146-163.

Doyle, P. (2000). *Value-based marketing: Marketing strategies for corporate growth and shareholder value*. Chichester: Wiley.

Fournier, S., Dobsha, S., & Mick, D.G. (1998). Preventing the premature death of relationship marketing. *Harvard Business Review,* 42-51.

Grover, V., & Teng, J.T.C. (2001). E-commerce and the information market. *Association for Computing Machinery, Communications of the ACM, 44*(4), 79-86.

Hagel III, J. (1999). Net gain: Expanding markets through virtual communities. *Journal of Interactive Marketing, 13*(1), 55-65.

Hagel III, J., & Rayport, J. (1997). The coming battle for customer information. *Harvard Business Review,*53-65.

Moon, Y. (2000). *Network technology and the role of intermediaries.* Boston: Harvard Business School.

O'Malley, L., Patterson, M., & Evans, M. (1997). Intimacy or intrusion? The privacy dilemma for relationship marketing in consumer markets. *Journal of Marketing Management, 13*(6), 541-559.

Phelps, J., Nowak, G., & Ferrell, E. (2000). Privacy concerns and consumer willingness to provide personal information. *Journal of Public Policy & Marketing, 19*, 27-41.

Van Raaij, W.F., & Poiesz, T. (2003). Rethinking the value concept in marketing. In P.J. Kitchen (Ed.), *The future of marketing; Critical 21st century perspectives*. New York: Palgrave Macmillan.

Wind, J., & Rangaswamy, (2001). Customerization: The next revolution in mass customization. *Journal of Interactive Marketing, 15*(1), 13-32.

Chapter XIX

Meeting the Demands of Wide Audience End Users

Ken Peffers, University of Nevada, USA

Tuure Tuunanen, Helsinki School Economics, Finland

Abstract

This chapter identifies seven problems associated with requirements elicitation, where the intended users are external to the firm; it proposes characteristics for requirements elicitation methods to resolve the problems and tells the story of a new method that supports the resolution of six of the seven problems. Diverse, external users with little relationship to the organization present special problems for requirements elicitation. We identify seven problems for requirements elicitation with wide audience end users (WAEU): context, reach, modeling, model aggregation, presentation, consensus building, and the requirements-design interface. We also identify seven characteristics for requirements elicitation methods to support solutions for these problems. We develop the critical success chains (CSC) method to support six of the seven characteristics and demonstrate its use to develop applications for mobile financial services at Digia, Inc, a Helsinki-based software development firm. Current research efforts are addressing the seventh problem.

Introduction

One of the key risks to the success of information systems (IS) projects is the failure to understand the functional requirements for the new systems (Lyytinen & Hirschheim, 1987). It is frequently claimed that the hardest single part of building a new IS is deciding precisely what features and functionality to build into the system, and no other part is as difficult to rectify later if it isn't done well (Brooks, 1975).

Requirements engineering (RE) is a process that involves all of the activities required to create and maintain system requirements that guide the information systems development work (Kotonya & Sommerville, 2002). Requirements elicitation is the initial phase in RE that focuses on collecting requirements of the would-be users. The traditional systems development view about requirements elicitation has been that requirements are out there somewhere to be gathered by the analysts (Kotonya & Sommerville, 2002). In this view, the problem of requirements elicitation is simply a matter of finding the right informants, using the right techniques, and then selecting, prioritizing, and agreeing on the requirements. Developers today face a variety of problems that make this traditional view less and less relevant.

More and more, end users are external to the developing organization, and would-be users are widely dispersed physically and geographically, making them difficult to reach (Tuunanen, 2003). Third-generation mobile software applications (Peffers, Gengler & Tuunanen, 2003) are typical examples. Such software is used by a very diverse set of users, some of them within the organization developing the software, but most in client organizations or with no connection to the developing organization. Understanding the needs of such users is a thorny problem (Salaway, 1987), and it is not made easier by the fact that many end users do not know how to express their needs (Walz, Elam & Curtis, 1993; Watson & Frolick, 1993). Consequently, it isn't easy to understand the needs of these users and exactly how to develop systems that meet those needs. Our objective is to develop research in order to develop methods for requirements elicitation to meet the needs of these users.

Here, we define a new type of system end user: the wide audience end user (WAEU). Many systems intended for WAEUs are emerging, such as embedded Java applications for 3G mobile phones or digital TVs for consumer markets. These systems are developed for end users who are not within an organization's reach, so traditional requirements elicitation methods are not well suited to support software engineers in approaching them. To understand the system requirements of WAEUs, we need to develop better ways to include them in the planning and development process. Otherwise, we risk facing problems that are all too familiar to developers of innovative products: development prototypes show great promise in the lab and among product engineers, but they are rejected by the intended customers when introduced to the markets, because they are too difficult to use or miss the mark of meeting the users' needs and desires (Peffers et al., 2003; Peffers & Tuunanen, 2005).

Key Characteristics for Wide Audience Requirements Elicitation

Reach and communication provide a useful starting point for understanding key characteristics of wide audience requirements elicitation (Tuunanen, 2003). We incorporate these two dimensions in our framework, as shown in Figure 1. This framework extends the work of Hickey and Davis (2003) and Nuseibeh and Easterbrook (2000) about requirements elicitation techniques and work on the communication aspects of elicitation (Curtis, Kellner & Over, 1992; Curtis, Krasner & Iscoe, 1988; Davidson, 2002; Keil & Carmel, 1995). The figure shows the three levels of reach, starting from the analyst-developer in the center and moving away from them to users that are successively harder to reach: user representatives, user groups, and the user community or WAEU. We use single- and two-directional arrows, suggesting the capabilities of categories for effec-

Figure 1. Techniques for requirements elicitation with wide audience end users suggesting the effectiveness of communication and reach (Tuunanen, 2003)

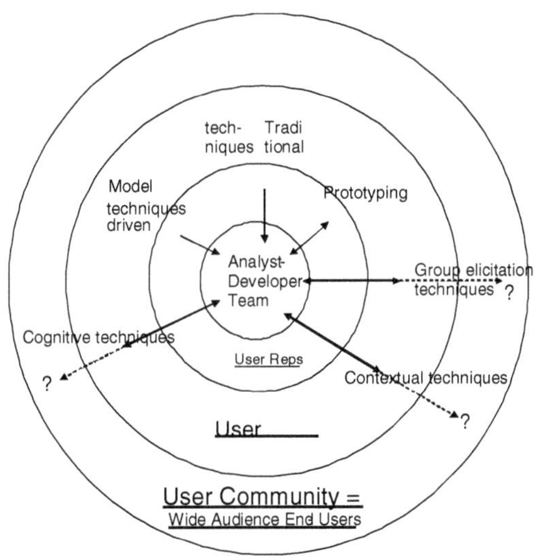

tive communication among the parties. This results in three proposed technique categories—group elicitation techniques, contextual techniques, and cognitive techniques—marked with question marks, because we think that techniques in these categories may be adapted to make them more effective for requirements elicitation with WAEU.

In our recent research (Peffers et al., 2003; Peffers & Tuunanen, 2005; Tuunanen, 2003; Tuunanen, Peffers & Gengler, 2004), we found seven distinct problems associated with requirements elicitation for WAEU (Tuunanen et al., 2004): context, reach, modeling, model aggregation, presentation, consensus building, and the requirements-design interface.

- **Context:** WAEUs often have little or no historical relationship with the firm, the product line, or the technology and, hence, may have little context in which to have ideas about desirable functionality (Salaway, 1987). This is particularly true when developers wish to design new applications with features hitherto unavailable (Peffers et al., 2003), often the case where the product is innovative.
- **Reach:** WAEUs cost more to reach for data collection than in-house users and are likely to be unavailable for iterative or interactive consultation about their needs. When we reviewed more than 100 techniques in our research, we found that most researchers assume that end users are available and that the analyst-developer team has control over them (Peffers et al., 2003; Peffers & Tuunanen, 2005; Tuunanen, 2003; Tuunanen et al., 2004). This is seldom the case for systems designed for external users.
- **Modeling:** The characteristic of user knowledge is likely to differ sufficiently from that of developers so much that it isn't easy for decision makers to understand what they want, why they want it, or the importance of their preferences (Watson & Frolick, 1993). This is not something new for developers, but it becomes more important as product life cycles are shortened, and applications using innovative technologies are put to market before the public has had a chance to become familiar with the underlying technologies.
- **Model Aggregation:** WAEU perspectives may differ sufficiently such that it becomes difficult to aggregate their preferences to present a meaningful, aggregated view for decision makers. Researchers trying to understand the needs of a consumer segment, for example, may conduct studies involving dozens or hundreds of individuals (Kotler, 1994). Individuals from various cultures, occupations, and interest groups are likely to hold perspectives sufficiently different so as to challenge analysts in meaningfully aggregating their views.

- **Presentation:** Differences in perspective and culture between WAEUs and managers may make it difficult for managers to understand and evaluate data from WAEUs to make decisions about which features to incorporate and how to do so. That top-down management can provide a solution (Shank, Boynton & Zmud, 1985), for this is a traditional mainstream view in IS, but executives may not know enough about what users need or want in order to fill this role. Scandinavian IS researchers Bjerknes and Bratteteig (1995) have argued for intensive participation by end users and the incorporation of end-user views into the development process in order to increase project success rates. This creates demands for presenting the information and communicating it to a variety of stakeholders. This is not a new problem for the IS developer, but it is exacerbated by the physical, conceptual, and cultural distance between the developers and the would-be users (Barki, Rivard & Talbot, 1993; Dennis, George, Jessup, Nunamaker Jr. & Vogel, 1988; Nunamaker, Dennis, Valacich, Vogel & George, 1991).
- **Consensus Building:** Managers may lack the concepts and tools necessary to make the most effective decisions about features and attributes, the source of which is external to the organization. The literature offers ways of reaching consensus (Davison & Briggs, 2000; Herlea & Greenberg, 1998; Linstone & Turoff, 1975; Nunamaker et al., 1991) and prioritizing requirements (Green & Krieger, 2001; Herzwurm, Schockert & Pietsch, 2003; Herzwurm, Schockert & Weinberger, 1997; Johnson, 1987; Laaksonen, Tuunanen & Rossi, 2004; Ravichandran & Rai, 1999, 2000; Zultner, 1993). However, for the time being, there is little literature available that connects the consensus-building approach to a more comprehensive approach (Tuunanen, 2003).
- **Requirements-Design Interface:** It may be hard to model the results of the whole requirements engineering process in forms that permit WAEU views to be used effectively in the design process. Furthermore, requirements elicitation techniques found in the literature are not widely used by practitioners (Nikula, Sajaniemi & Kälviäinen, 2000a, 2000b). Ideally, the results of the requirements engineering process should be in a form that can be readily taken up by system designers. This idea has been the subject of some discussion in the research community (Briggs & Gruenbacher, 2002). It would be particularly valuable if the RE outcome were consistent with software engineering standards (e.g., *IEEE Standards Software Engineering, Volume Four: Resource and Technique Standards*, 1999) and could be used directly by tools that are embedded within the development environment (Ramesh & Jarke, 2001).

Table 1. Key characteristics for requirements elicitation technique

1.	**Context.** Data gathering method that does not require users to understand firm or technology.
2.	**Reach.** Data sufficiently rich so that interaction is not required. Data gathering is economical.
3.	**Modeling.** Modeling user preferences and values flexibly.
4.	**Model Aggregation.** Aggregating user ideas quickly and flexibly.
5.	**Presentation.** Ability for developers to easily examine data at different levels of aggregation.
6.	**Consensus Building.** Supporting consensus-reaching behavior.
7.	**Requirements-Design Interface.** Present models of new features and attributes in a semi-structured form useful for systems design.

These seven problems for requirements elicitation with WAEUs can form the basis for the means to addresses all of them. These requirements are summarized in Table 1. An RE method that incorporates these characteristics might be expected to effectively support requirements elicitation from WAEU.

Illustrating the Potential Technique

How might these seven characteristics be supported in one technique? Here, as an example, we discuss our efforts with Digia, Inc., a leading Symbian mobile software developer in Finland,[1] to develop applications for mobile financial services applications (Peffers & Tuunanen, 2005). We used the critical success chains (CSC) method to elicit requirements in this case. Critical success chains method is a data-gathering and analysis method for IS planning (Peffers & Tuunanen, 2005). The case illustrates the potential of CSC for WAEU requirements elicitation.

Context and Reach

To reach people, we first had to select the study participants. Our mission was to develop ideas for new applications in mobile financial services, ideally locating the potential killer application for the financial sector with third-generation mobile terminals. We wanted to obtain input from people who might be expected to possess knowledge about the value of potential new applications, regardless of their role up and down the value chain. In this project, we chose people who worked as bankers, risk investors, mobile telecommunications operators, researchers, and potential end users.

Table 2. Participant selection for Digia case study (Peffers & Tuunanen, 2004)

THE BUSINESS USER SEGMENT OF MOBILE VALUE-ADDED SERVICES	
Description	Aware and has used mobile value-added services (VAS)
Wealth	Not restrained by monetary issues
Education	Highly educated, working on the managerial level or as an expert
Technology orientation	Internet and information technology littermates
Mobility	Very mobile people, participating actively in sports and outdoor activities
SMS usage	Moderate SMS users (1-7 mobile originated SMS/week)

We focused on two groups of participants—experts and end users. For potential expert participants, we targeted a cross-section of Finland's most relevantly knowledgeable scientists, professionals, and managers. Our analyst worked with Digia's chairman and a staff assistant to develop a list of expert participants. For end users, we targeted sophisticated communication technology users, people who likely would be early adopters of attractive new mobile commerce applications or so-called lead users (von Hippel, 1986). We defined the desired demographic characteristics for end-user participants using segmentation that had been used in a major mobile value-added services study (Nokia, 1999). For the purposes of this study, the business mobile user segment, described in Table 2, was considered the most appropriate.

The selection of potential early adopters was more complex. We wanted to identify potential early adopters for applications that hadn't yet been invented. We used the snowball method to identify professional, managerial, executive, and other sophisticated end users with the characteristics of the business mobile user segment, starting with individuals identified by Helsinki School of Economics faculty members. The resulting combined list of nominated participants contained 40 names. Snowballing refers to asking well-known lead users to recommend another one, until we have enough leads. After selection process, we wanted collect idea stimulus for the context area. We did this by asking each participant during the appointment-setting telephone call to tell us the first thing that came to their minds when they thought about financial services that could be provided with next-generation mobile devices. Thirty of the 32 participants provided us with their ideas. We rewrote the 30 ideas as four system descriptions from which participants could infer specific features.

Modeling Requirements

In the Digia study, we interviewed each of the participants individually over a five-week period. The interviews usually averaged from 45 to 60 minutes. In each interview, we spent the first five to 10 minutes on warmup conversation and

Figure 2. Example chain collected from a participant in the Digia study

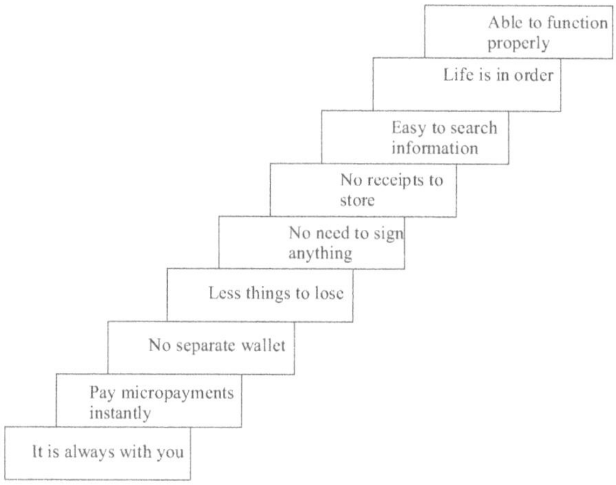

discussion of the interview objectives. Then, we showed the participant the four system descriptions and asked him or her to rank order the best two. Generally, the participants volunteered ideas about system features. When necessary, we also showed participants prototype pictures of innovative mobile devices to indicate that we weren't limited by current device capabilities.

Next, we asked the participant a series of questions to collect chains of attributes, consequences, and personal objectives (Peffers, Gengler & Tuunanen, 2003). All but five of the interviews were done at the participant's work premises, because participants seemed most at ease there, especially when it was in a conference room away from interruptions rather than in the participant's office. The interviews were recorded.

Figure 2 shows an example chain. Items near the bottom refer to an application's specific attributes. Those in the middle refer to consequences or performance. Those near the top refer to results, objectives, or values. More specific information about the interview technique can be found in Peffers and Gengler (2003), Peffers et al. (2003), and Peffers and Tuunanen (2005).

Requirements Model Aggregation

The data can easily contain hundreds or even thousands of distinct statements. This makes it difficult, if not impossible, for decision makers and designers to interpret directly without analysis. It was important to aggregate this data to produce a meaningful yet smaller set of rich, unified aggregated models that managers and designers then could grasp.

The objective was to create at the top layer an aggregated representation of participant models and semantic maps (Peffers & Gengler, 2003; Peffers et al., 2003; Peffers & Tuunanen, 2002). We clustered the chains using Ward's method in order to minimize the variance of the constructs contained in each cluster, aggregating the chains into socially constructed critical success models (Aldenderfer & Blashfield, 1984). We examined resulting clusters, focusing on solutions with three to 10 clusters. After producing rough models for most of the cluster solutions, we settled on a five-cluster solution that seemed to be coherent and meaningful. We mapped each cluster onto a network model, where nodes represent the constructs (attributes, consequences, and values/goals), and links represent the links connecting the constructs into meta-level chains.

Presenting the Requirements

The next phase in the process was to transform the clustered chains into a network map. These maps contain features (application attributes), reasons why customers saw them necessary or interesting (critical success factors), and, finally, personal goals or values driving the customers. Figure 3 shows an example network model from the Digia study, which describes a mobile wallet application concept. Each node in the figure represents a statement by study participants. The number in the circle indicates the number of participants who voiced this statement.

Figure 3. Mobile wallet map from Digia study

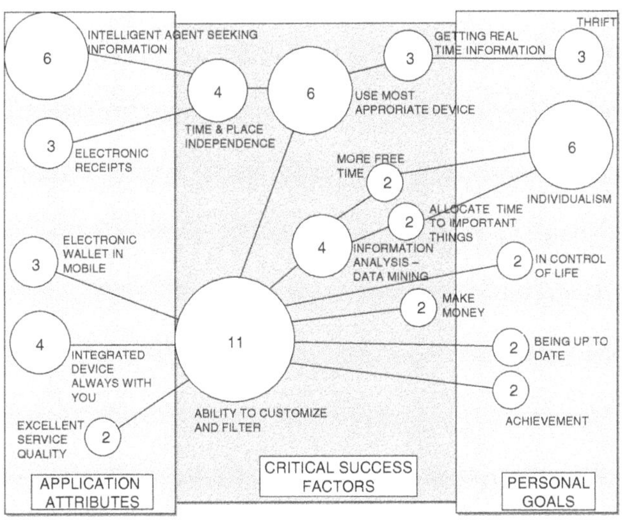

Consensus Building

In the final phase, these graphical presentations, or maps, were presented in an initial workshop in order to introduce the key features of the system to the client's R&D people, who evaluate the maps and identify the feasible project ideas. With the Digia case, both business and technical R&D people were chosen to participate, including the chairman of the board, the Nokia key account director, two business development managers, and two engineering managers.

The workshop was scheduled for a single five-hour session, starting with a brief introduction to the CSC method and the study and a discussion of the purpose of the meeting. The objective was to examine and discuss each CSC map as a group and then come up with a project idea that would address the desired consequences and values expressed in the models, including a name, short description, architecture, a list of supply-chain players, customer segment(s), benefits for players and customers, profit model, and risks involved, all at a back-of-the-envelope level of detail. In addition, the goal was to produce a rough graphical network business model that showed the relationships among customers and supply chain participants, including the flow of information, value, and revenue. Participants were told that they should rely on their own knowledge rather than making use of resources outside the room.

For each of the CSC models for which an idea was developed, the participants first drew a business model on the flipchart. These we further developed in a post-workshop analysis. The main focus was in proof-drawing the models and providing a coherent way to express the resulted information. Finally, as a final product, we provided the client a business report that summarized the findings.

According to Digia product analyst Markus Ahonen, the firm regards CSC favorably because (1) the method seems to work well, and it is easy to see why it works; (2) the interviewer collecting data for the method needs no special abilities or skills; and (3) the resulting CSC maps are very helpful in understanding how people think about an issue. Digia's chairman remarked that the workshop "positively...exceeded [his] expectations [about] the results..." (Peffers & Tuuanen, 2005). The firm plans to continue using the CSC method in IS planning (Peffers & Tuunanen, 2005).

Conclusion

We have presented seven characteristics for elicitation techniques that we see as important in eliciting system requirements for WAEUs. We showed through a demonstration study, using the CSC method, that most of these characteristics can be supported in the RE process. By using laddering (Peffers & Tuunanen, 2005) and the lead-user concept (von Hippel, 1986) CSC overcame problems with context and reach. We find that laddering technique that have been incorporated in CSC (i.e., modeling requirements) also provides us with the means to model the requirements flexibly without rigid preconceptions and in a way that easily can be aggregated. We suggest using clustering for this task (Aldenderfer & Blashfield, 1984).

We recognize the lack of requirements-design interface. In our current research, we are extending the method to resolve this (Tuunanen et al., 2004) by extending the new methods so that the outcome of requirements elicitation can be directly used in design. In working on these issues, we seek to conduct rigorous research that has practical relevance.

Endnote

1 Further details available at www.digia.com or (Peffers & Tuunanen, 2005)

References

Aldenderfer, M.S., & Blashfield, R.K. (1984). *Cluster analysis*. Beverly Hills: Sage.

Barki, H., Rivard, S., & Talbot, J. (1993). Toward an assessment of software development risk. *Journal of Management Information Systems, 10*(2), 203-225.

Bjerknes, G., & Bratteteig, T. (1995). User participation and democracy: A discussion of Scandinavian research on system development. *Scandinavian Journal of Information Systems, 7*(1), 73-98.

Briggs, R.O., & Gruenbacher, P. (2002). EasyWinWin: Managing complexity in requirements negotiation with GSS. *Proceedings of the 35th HICSS*, Kona, HI.

Brooks, F. (1975). *The mythical man month: Essays on software engineering*. Reading, MA: Addison-Wesley.

Curtis, B., Kellner, M.I., & Over, J. (1992). Process modeling. *Communications of the ACM, 35*(9), 75-90.

Curtis, B., Krasner, H., & Iscoe, N. (1988). A field study of the software design process for large systems. *Communications of the ACM, 31*(11), 1268-1287.

Davidson, E.J. (2002). Technology frames and framing: A socio-cognitive investigation of requirements determination. *MIS Quarterly, 26*(4), 329-358.

Davison, R., & Briggs, R.O. (2000). GSS for presentation support. *Communications of the ACM, 43*(9), 91-97.

Dennis, A.R., George, J.F., Jessup, L.M., Nunamaker Jr., J.F., & Vogel, D.R. (1988). Information technology to support electronic meetings. *MIS Quarterly, 12*(4), 591-624.

Green, P.E., & Krieger, A.M. (2001). Thirty years of conjoint analysis: Reflections and prospects. *Interfaces, 31*(3), S56-S73.

Griffin, A., & Hauser, J.R. (1993). The voice of the customer. *Marketing Science, 12*(1), 1-27.

Herlea, D., & Greenberg, S. (1998). *Using a groupware space for distributed requirements engineering* [technical report No. 98/614/05]. Calgary: University of Calgary, Canada.

Herzwurm, G., Schockert, S., & Pietsch, W. (2003). QFD for customer-focused requirements engineering. *Proceedings of the 11th International IEEE Requirements Engineering Conference*, Monterey, California.

Herzwurm, G., Schockert, S., & Weinberger, C. (1997). Customer oriented evaluation of QFD-tools. *Proceedings of the Third International Symposium on Quality Function Deployment*, Linköping, Sweden.

Hickey, A.M., & Davis, A.M. (2003). Requirements elicitation and elicitation technique selection: A model for two knowledge-intensive software development processes. *Proceedings of the Hawaii International Conference on Systems Sciences*, Hawaii.

IEEE standards software engineering, volume four: Resource and technique standards [No. IEEE Std. 830-1998 (Revision of IEEE 830-1993)]. (1999). IEEE.

Johnson, R.M. (1987). Adaptive conjoint analysis. *Proceedings of the Sawtooth Software Conference,* Ketchum, ID.

Keil, M., & Carmel, E. (1995). Customer developer links in software-development. *Communications of the ACM, 38*(5), 33-44.

Kotler, P. (1994). *Management analysis, planning, implementation, and control*. Englewood Cliffs, NJ: Prentice-Hall.

Kotonya, G., & Sommerville, I. (2002). *Requirements engineering, processes and techniques*. New York: John Wiley.

Laaksonen, A., Tuunanen, T., & Rossi, M. (2004). Requirements validation for consumer software by conjoint analysis. *Proceedings of the ISD 2004*, Riga, Lithuania.

Linstone, H.A., & Turoff, M. (Eds.). (1975). *The Delphi method techniques and applications.* MA: Addison-Wesley.

Lyytinen, K., & Hirschheim, R. (1987). Information-systems failures—A survey and classification of the empirical literature. *Oxford Surveys in Information Technology, 4*, 257-309.

Nikula, U., Sajaniemi, J., & Kälviäinen, H. (2000a). Management view on current requirements engineering practices in small and medium enterprises. *Proceedings of the Fifth Australian Workshop on Requirements Engineering*, Brisbane, Australia.

Nikula, U., Sajaniemi, J., & Kälviäinen, H. (2000b). A state-of-the-practice survey on requirements engineering: Industry expectations on education and technology transfer. *Proceedings of the Fifth International Conference on Software Process Improvement Research, Education and Training—INSPIRE V*, London.

Nokia. (1999). *The demand for mobile value added services.* Study of Smart Messaging, Market Study. White paper. Available online *www.nokia.com*

Nunamaker, J.F., Dennis, A.R., Valacich, J.S., Vogel, D., & George, J.F. (1991). Electronic meeting systems to support group work. *Communications of the ACM, 34*(7), 40-61.

Nuseibeh, B., & Easterbrook, S. (2000). Requirements engineering: A roadmap. *Proceedings of the Future of Software Engineering, ICSE 2000*, Limerick, Ireland.

Peffers, K., & Gengler, C. (2003). How to identify new high-payoff information systems for the organization. *Communications of the ACM, 1,* 83-88.

Peffers, K., Gengler, C., & Tuunanen, T. (2003). Extending critical success factors methodology to facilitate broadly participative information systems planning. *Journal of Management Information Systems, 20*(1), 51-85.

Peffers, K., & Gengler, C.E. (2003). How to identify new high-payoff information systems for the organization. *Communications of the ACM, 46*(1), 83-88.

Peffers, K., Gengler, C.E., & Tuunanen, T. (2003). Extending critical success factors methodology to facilitate broadly participative information systems planning. *Journal of Management Information Systems, 20*(1), 51-85.

Peffers, K., & Tuunanen, T. (2002). Using rich information to plan mobile financial services applications with maximum positive impact: A case study. *Proceedings of the Mobile Round Table*, Tokyo.

Peffers, K., & Tuunanen, T. (2005). Planning for IS applications: A practical, information theoretical method and case study in mobile financial services. *Information & Management, 42*(3), 483-501.

Ramesh, B., & Jarke, M. (2001). Towards reference models for requirements traceability. *IEEE Transactions on Software Engineering, 27*(1), 58-93.

Ravichandran, T., & Rai, A. (1999). Total quality management in information systems development: Key constructs and relationships. *Journal of Management Information Systems, 16*(3), 119-155.

Ravichandran, T., & Rai, A. (2000). Quality management in systems development: An organizational system perspective. *MIS Quarterly, 24*(3), 381-415.

Salaway, G. (1987). An organizational learning approach to information systems development. *MIS Quarterly, 20*(1), 244-264.

Shank, M.E., Boynton, A.C., & Zmud, R.W. (1985). Critical success factor analysis as a methodology for MIS planning. *MIS Quarterly, 9*(2), 121-129.

Tuunanen, T. (2003). A new perspective on requirements elicitation methods. *JITTA: Journal of Information Technology Theory & Application, 5*(3), 45-62.

Tuunanen, T., Peffers, K., & Gengler, C.E. (2004). *Wide audience requirements engineering (WARE): A practical method and case study*. Helsinki: Helsinki School of Economics.

Walz, D., Elam, J., & Curtis, B. (1993). Inside a software design team: Knowledge acquisition, sharing and integration. *Communications of the ACM, 36*(10), 62-77.

Watson, H.J., & Frolick, M.N. (1993). Determining information requirements for an eis. *MIS Quarterly, 17*(3), 255-269.

von Hippel, E. (1986). Lead users: A source of novel product concepts. *Management Science, 32*(7), 791-805.

Zultner, R.E. (1993). TQM for technical teams. *Communications of the ACM, 36*(10), 79-91.

Chapter XX

Envisioning the Future of a Multi-Channel World By 2020

Markku Tinnilä, Helsinki School of Economics, Finland

Theresa Lauraeus-Niinivaara, Helsinki School of Economics, Finland

Abstract

Envisioning the future is always challenging, and the multi-channel environment is no exception to that rule. The objective of this study was to hypothesize upon the kind of digital services that will likely develop by the year 2020 in a networked, multi-channel business environment. Our starting point was to approach the analysis from a business perspective; in other words, we tried to avoid the typical technology-focused e-business forecasting, which charts the development of new technologies and devices. Instead, we tried to survey for services that would be available across different electronic channels and meet customer needs in the future, as well as what types of businesses would develop around these services. Consequently, this study gives answers to questions like, what will be the future of e-services, who needs and uses them, how will the customer want to buy such services, who will produce e-services and what will the earnings logic be based upon?

Introduction

Envisioning the future is always challenging, and the multi-channel environment is no exception to that rule. We all remember the brave new world predicted by the hype years of e-business at the turn of the millennium. Few of those visions have been realized thus far. However, as the doyen of science fiction writers, Sir Arthur C. Clarke, has stated in his Third Law, "The only way of discovering the limits of the possible is to venture a little way past them into the impossible." Thus, we have to create visions to weed out the unrealistic predictions and focus on the development of the feasible services.

The vision presented here is based on a study of expert and layman opinions, evaluated by the same group. The objective of this study was to hypothesize upon the kind of digital services that will likely develop by the year 2020 in a networked, multi-channel business environment. Our starting point was to approach the analysis from a business perspective; in other words, we tried to avoid the typical technology-focused e-business forecasting, which charts the development of new technologies and devices. Instead, we tried to survey for services that would be available across different electronic channels and meet customer needs in the future, as well as what types of businesses would develop around these services. Consequently, this study gives answers to questions like, what will be the future of e-services, who needs and uses them, how will the customer want to buy such services, who will produce e-services and what will the earnings logic be based upon?

The time-perspective of the study was set at year 2020. With the year 2020, we tried to reach far enough into the future but, at the same time, keep our feet on the ground. Charting the development of the next few years is usually an extrapolation of the present day, based on development of today's technology, whereas, when you are trying to predict far ahead into the future, you very often make assumptions that technological development has resolved all the problems of today. The choice of the year 2020 hopefully balances both of these perspectives.

The goal was to create visions that could be taken into consideration in long-term strategic planning. Although some of the introduced ideas and visions are far from the current situation, we have to remember that only a dozen years ago, the very idea of a worldwide electronic network available for everyone was completely alien. This clearly has been the case with the Internet, mobile phones, and digital television. Internet and mobile services have reached great penetration and hundreds of millions of users within a decade, a time period much shorter than from today to 2020. With this backward glance to history, we want to point out that the visions presented should be viewed in proper perspective and need to be gauged for their feasibility in each industry and business.

The visions have been produced using an Internet-based Weak Signals Tool. With it, we have collected more than 200 ideas and signals from different expert groups and have had 87 experts generate ideas. This chapter condenses the results of several research papers dealing with the same topic (Lauraeus & Tinnilä, 2003; Tinnilä & Lauraeus, 2004).

Envisioning the Future of E-Business

Many people are interested in getting a glimpse into the future, so there are dozens of experts offering their personal views on it. In this chapter, the study of the future is based not only on the interviews of a few experts and individual opinions, but on a consensus view of a group of experts. The empirical study was carried out in Finland with the help of the Weak Signals Tool developed by Fountain Park Ltd (Ilmola et al., 2002). The tool is an Internet-based platform, where one basic question or theme is the starting point. In this case, we asked, "What kind of digital services will develop by the year 2020 to the networked multi-channel environment?" Based on this theme, the participants wrote down their ideas in the Internet form. The ideas were stored, and, in the second phase, the same group of participants evaluated the ideas. Each participant was provided randomly with 20 ideas for evaluation. The evaluation was done by positioning the idea for relevance. The closer to the center, the more relevant the idea was graded. After 20 evaluated ideas, a new set of ideas was given, and this was repeated as long as the participants were willing. This method provided us with more than 200 ideas for rating. The participants were university researchers, business and development managers, and business students.

The tool divides the answers presented into four different categories, called here the dominant view, potentially important factors, unimportant factors, and weak signals. The dominant view consists of answers in which all the contributing experts share the same view with regard to their importance and relevance. The potentially important factors are those views that are also relevant but have a higher deviation of opinions among them. The unimportant factors are those where all agree on their low relevance. The last but not the least group of views consists of those ideas that only a few experts recognize as important. These so-called weak signals, however, are those that may have the greatest influence. Obviously, the lines between these groups are not static. Potentially important factors may become the dominant view, when they are generally agreed on. Similarly, the weak signals seen by only a few may prove to be the surprising success factors in the future. We start by presenting the dominant view, where there is the greatest consensus of opinion.

Figure 1. The weak signals tool (adapted from Fountain Park Ltd.)

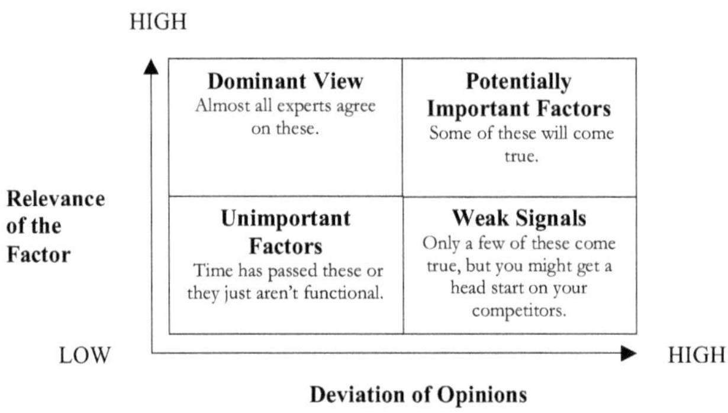

Dominant View

The dominant view consists of signals that have been ranked to be the most relevant and have the least dispersion of opinion among the participants of the study. Explanations and researchers' interpretations have been added to these signals. The results can be found in their original form in Lauraeus and Tinnilä (2003). The dominant views can be described as follows:

Improving Efficiency in B2B Business

The improved efficiency of businesses through B2B services will take place via machine-to-machine (M2M) integration such as Web services. With M2M interaction, it is possible to get more efficient operations than in the first generation of electronic commerce when human beings are using the systems. This M2M integration will automate many thus far manual activities (e.g., in purchasing, production line replenishment, etc.).

In B2C commerce, there will be an increase in the role of communities. This need for new types of communities is manifested in always-online communication. Information channels will be accessible by every member of a community all the time and from anyplace.

The Consumer Chooses the Best Services with His or Her Wallet

Today, everyone can produce his or her own Web pages. In the future, everyone will be able to produce digital services (e.g., video clips, music, pictures).

Determining which e-services will survive into 2020 is simple: the consumer will vote for viable solutions with his or her wallet. The trend of everyone producing his or her own content is already visible on the Internet and in digital television reality programs. In both cases, it is presumed that the users take part in content production. The consumers choose those that are most suitable for their needs from these individually created services as well as commercially produced services.

Something for Everyone

Electronic services and content will provide something for everyone. The data terminal or access device is only one dimension of service, unlike today where the access device (e.g., mobile phone) determines what services can be used. Bank services are a good example; branch offices, Internet and mobile, all provide a widely different set of services. The direction of development is that, in the future, all services can be used, regardless of access device. Multi-channel environment allows all customers and consumers to have their own preference in selecting the desired terminal.

Versatile Usage Possibilities

The architecture solutions will develop in the direction that e-services will become terminal independent. Everyone can use services from a terminal suitable to him or her at any time. This requires from services a more versatile structure and for conversion services to provide the desired range of services and content to each access device.

Services Everywhere and Anywhere

The wireless networks and the portable data terminals make it possible to use electronic services anywhere. As mentioned in two of the earlier signals, the emphasis is on the possibility to use e-services time and place independently. In this signal, experts emphasized independence from fixed networks as the way of the future, unlike present wireless systems that have capacity and terminal limitations.

Communication and Entertainment for Individuals

Future services combine entertainment and communication services, tailored to individual needs and tastes. Personal preferences in terms of content can be provided. Some people will want online news, while others may prefer entertainment, and still others, e-communities. When used by a critical mass of individuals, these services also come commercially viable. This signal emphasizes individuality and freedom of choice. This trend already has been clear in the Internet services; however, the volumes so far have been insufficient to support a large range of commercially viable services.

Customer Power and Knowledge Increases

In the future, our experts predict that the customer's power (i.e., freedom of choice) and knowledge of products will increase. Several studies indicate that the majority of people already use Internet as a primary channel when reaching information about products and services. The customer's power and choice allowance will increase, when markets become more transparent. Examples of services enabling this are net-based price agents that can be used to compare prices, delivery terms, and so forth, of all kinds of services from travel to electronics.

Ease of Use: A Key Factor in New Services

Non-English-speaking children learn to use computer games by playing and experimenting with them. Why does this not apply to other e-services? Why can't services be designed in a way that facilitates their use without prior reading of manuals? The reason behind the failure of many mobile and Internet-based services is due to users' difficulties when using the services for the first time. The first impression is difficult to change, and if it does not work on the first try, consumers often abandon the service. The logic behind these opinions is the need to develop standardized ways of using services. People do not want to spend time learning how to use each service separately. Consumers want to have easy-to-use, entertaining, and quick solutions to serve their different needs.

Service Availability at Home Whenever I Need It

Experts predict that all the services people need will be available in their homes by the year 2020. People won't have to leave their houses to visit shops, manage finances, reserve travel tickets, or even complain about taxes. The first groups needing this type of services are likely disabled and elderly people. For example, the public library is a nice place to visit, but the books reserved from the Internet also can be delivered to people's doors. This type of environment also facilitates remote telecommuting. The major obstacles to realizing this environment are practical problems. The technology is already available, but, in many cases, there are problems connected with identification, methods for ease of paying, trust, and reliability of different channels. In many cases, the solutions, for example, are reliable identification, on hand, but the lack of standardization has prevented their large-scale adoption.

Potentially Important Factors

Potentially important factors provide an alternative outlook on the future. These factors have been estimated to be potentially important, but experts' opinions differ about their significance. Some have ranked these as being very important, while others have not recognized their value. Therefore, there is less consensus on these ideas, and they may provide for surprises in the future. The potentially important factors are described as follows:

The Expert is the King

Instead of a leading standard of tools, experts in different fields likely will be the main actors in the future. Electronic systems are like a bottle of milk in that everyone can pour from it. The contents are a separate matter (e.g., lactose-free milk). The idea behind the increasing importance of experts is that electronic systems enable everyone to do the basic things easily. However, only experts are able to create really new content and services.

General Entertainment and Work Devices are on their Way

In the long run, people likely will begin to move their work from their offices to their homes. If that happens, joint entertainment and work centers may be required. This center will be the new most important device at home, and we predict that people will be willing to spend a lot of money on it. The convergence of entertainment devices may show the way in creating these devices. Instead of separate videos, game devices, televisions, and stereos, there will be multi-function devices.

Communication is Cheap or Even Free

Communication and automated services are expected to be free or at least very inexpensive in the future. This means that consumers will not be willing to pay for expensive services. This is obvious even now, as news, for example, can be retrieved free from the Internet, and many ideas based on selling paid contents have failed. The exceptions include services where the payer, for example, is the employer on behalf of employees. This also may mean that automatic network-based services may be priced on a much lower scale than personalized services. For example, banks currently offer both types of services—the automatic teller machine and Internet services—which are either free or cheap, and branch office-based personalized services, which are available only at a premium price or to key customers.

Consumers Want to Know What They Are Buying

With many of the current electronic service offerings, it is unclear for the customer what he or she is actually paying for. As consumers are unwilling to buy content or services that are found free elsewhere, the commercial services need to be differentiated. In many cases, the service price doesn't seem to have any relation to the core service provided. For example, why would anybody want to pay more to be able to pay for parking with a mobile phone, if it really is easier to use coins?

Shopping Remains Popular (Some Matters Never Change)

A lot of people will continue to want to shop for physical goods in stores (i.e., to go in to different shops, to touch, to look, to try on clothes, to test products, etc.). Consequently, multi-channel environment means that besides digital channels, there are traditional channels, which will continue to have their place in the future.

Timing is Important: Having the Right Services at the Right Place and Time

If the right services or products are provided at the appropriate environment at the right time, things start to grow. This adage applies to business as well as to gardening. The challenge for management is in having the right and accurate information of the customer behavior and needs, and an understanding of the coming trends and fashions. Many innovative services have failed in mobile, digital television, and Internet due to being too early.

Every Company and Community Can Produce Electronic Services

The infrastructure required for the production of e-services will be available to everyone and, as a consequence, costs will fall. Thus, every company or public organization will be able to produce electronic services just as easily as they can build their own Web pages now.

Internet Will be with You Everywhere and in the Size of a Sheet of Paper

Mobile displays and screens will be available in letter size and will be as thin and light as a piece of cardboard. In the future, telephone and its screen will become ordinary entertainment channels providing access to the Internet, games, and news, while users commute by underground subways or by bus. The electronic services will become part of the basic infrastructure of the society in the same way as use of paper, electricity or water.

The Bundling of E-Services Will Enable Operational Efficiency

In the future, one of the main problems will be how consumers and relevant services can find each other. This will be achieved by cooperation and networking in service delivery. The new bundled services will provide everyone with a suitable range of services. The finding and bundling of services requires players focused on providing these services. These infomediaries and servicemediaries base their business on providing consumers with easily found and bundled service packages, and content and service providers the access to large customer segments.

The Weak Signals: The Wild E-Cards

Obviously, while it is important to be familiar with the dominant view, it is hard to gain a competitive edge by doing the things everyone already regards as the industry standard. People who are able to exploit great new business opportunities are the ones who can recognize weak signals and take advantage of them, while others are still continuing the practices of yesterday. This story can be found in the rise of major players of current leading industries. One of the key elements behind the transformation of Microsoft from a small garage company into one of the largest companies in the world was the idea that in the future the software, and not the hardware, would be the bottleneck and, therefore, the key driver for adoption of computer use. IBM, as well as other big firms, did not fully appreciate this change before it was too late.

The weak signals present the greatest potential for surprises. They trace the very first hints and indicators of the things to come. Consequently, their importance is only understood by a few. These phenomena have no history, trend, or other past to be distinctly identified, but they still have the potential to become the focal points of new development. Therein lies the great interest in trying to catch these elusive signals; strategic advantage can be obtained by being the first to develop and offer a product or service. Business history is full of stories based on first-mover advantage, and many, if not most, large companies can trace the foundations of their success to similar innovations. Consequently, by analyzing weak signals, it is possible to recognize new ideas; innovators will be able to get ahead before their competitors, who rely only on past experience and trends.

If we were good at identifying the weak signals, we should able to recognize several of them at this very moment. Several years ago, the commercial potential of mobile services, global electronic networks, or content was understood by only a minority. Now, it is the dominant view; every layman on the streets is of the same opinion. The trick lies in being among the first to see the vital signals; weak signals live only for a moment. After that, they become more important and known by a large amount of people, or they just fade out.

The weak signals are part of the tacit knowledge found in every organization. However, this quiet information is poorly used by the organizations, and weak signals have many obstacles in their way on the road from observing them to understanding their significance. We tend to filter our observations according to what we consider reasonable. Weak signals often are turned down, as they appear at such an early stage that they do not yet have reasonable grounds. Still, everyone in any organization should look for signs of change. The earliest signs always are found at an individual level. When the tacit knowledge of the weak signals has been collected from company employees, it is up to the managers to make decisions based on them. To be able to recognize the relevant weak signals from the irrelevant ones, one needs to be familiar with the company's strategy and objectives (Ilmola et al., 2002).

Weak Signals in the Development of Technology and New Electronic Services

There are plenty of visions created on the technology of the future. In this study, we focus on trying to obtain foresight into the services and contents that customers and consumer can use through different digital channels. This section presents a group of weak signals collected in this research area, focusing on e-services in a multi-channel environment. The main weak signals are:

- **The electronic life control is possibly terminal independent: All different channels, smart objects and systems work smoothly together.** Only 10% of technologies that will form the basis for the solutions in the year 2020 exist at the moment. New innovations will have a significant role; for example, thinking microchips will be obtained with the help of DNA chains.
- **The creation of electronic services will not require massive information system projects anymore.** There will be out-of-the-box solutions available that will be easy and quick to implement.
- **The electronic expert services are breaking through.** All the essential information for company operations is in electronic form. This will make it possible for a company to obtain expert advice in small portions, so that even small companies can afford to utilize the supply of expert services. Instead of the tools, the power will lie with the experts. Tools have become relatively inexpensive and accessible to everyone so that they aren't relevant anymore, but, instead, content and product experts have become important actors.
- **Search services will be ubiquitously available.** Electronic services, no matter how they have been carried out in 2020, contain the answers to all our needs. The search engines will contain everything—offers, purchases, contacts, agreements, backups, information retrieval and social contacts
- **GPS and locator devices will be utilized in many matters.** For example, information will be retrieved by telephone by asking where the nearest restaurant with the cheapest lunch prices is.
- **Virtual reality will become an increasingly important part of our lives.** Soon, it will become more important than a physical reality. Already, there exist services on the Internet such as a virtual Helsinki, where users

can walk through virtual streets of the city, including all of the outskirts of a town. In the future, it will be possible to connect a physical reality to virtual reality with the help of satellite locators, cameras, measuring devices, and databases.

- **At home, a person can feel safe.** A person can do work and shop from home. Children can study with the help of new technology. To ensure safety and to protect homes, a complex alarm system will be built with the use of smart cards, access control systems, and television cameras. Domestic appliances will be integrated as part of a data network. The telephone will become a terminal for home computers by which users can switch the lights on, sound the alarm, turn the electricity on, heat up the stove, or turn on the vacuum cleaner robot. The portable computer will be integrated to the telephone so that it will be possible to read and write documents and to download and return documents into a company's main document control server.

Weak Signals Related to the Multi-Channel Network Economy

According to Saarinen and Tinnilä (2002), the typical features of the multi-channelnetwork economies are multiple channels, outsourcing, networking, and specialization in the network. Technological development has brought and is still bringing to the markets different access technologies that help customers to use the new services that companies are offering.

In the multi-channel environment, the production of the services through the cooperation of several companies has to be taken into consideration. Instead of emphasizing only electronic channels, attention also must be paid to the fact that different customer segments have different needs. The services and e-commerce must adapt themselves to the customer-desired data terminal equipment, connection channel, and operating situation. With the help of new technologies, it is possible to offer individual, customized services (Saarinen et al., 2002). These weak signals related to the multi-channel environment are:

- **In principle, the channel supply will increase. On the other hand, users' habits and preferences will dictate their utilization rate and, as a result, their development and lifespan, as well.** It is assumed that, in the future, users will try a lot of different alternatives, but at the same time, one of the alternatives will become the most commonly used standard, or a de facto standard.

- **The use of mobile service will be limited by the clumsiness of the user interface: A small display and keyboard won't be a good starting point.** The digital television can be suitable for impulse purchases, but the PC probably will remain the main device for the use of electronic services, as well, in the future.
- **In the information society, everyone has the double role of producer and user of information.** The digital television will be the channel that reaches even those that do not use the Internet or mobile phone. It will be essential for other groups, because it will be easy and quick to open and will be at home.
- **In the future, devices will be divided into two groups: fixed and mobile.** There always will be fixed devices at home, at work, and elsewhere, and then there will be mobile devices to be used outside in cars, in underground subways, and in the yard. Services and contents will have to be adapted to these two device groups. First, group devices have big screens and fixed locations, so their services can be more complicated and take advantage of the static user. Devices in the second group might be used anywhere, so usability and ease of use are the primary factors for service development. Mobile devices also will be used when users want information fast, whereas fixed devices will be used when users want the services in a wider scope.
- **Internet displays can be found anywhere; for example, in the backs of the seats of public buses.** The supply of information is impossibly huge, so extensive search services to hone in on the information needed will be important.

Conclusion

We have presented three different but converging views of future multi-channel e-business services in 2020. Obviously, it is not a complete picture and resembles more of a jigsaw puzzle, where only some of the pieces have found their places. Of the three views, the dominant view is the most commonly accepted. It already has a lot of consistent empirical evidence and is a logical trend to be followed. Investments to be carried out, based on the dominant view, thus can be estimated to within a reasonable limit, and profit calculations may even be possible. In this case, the future business and technology will be extensions of the present. The disadvantages connected to actions based on the dominant views are that no greater strategic advantages will be probable. When everyone has access to the same information, the markets are flooded with competitors with the same ideas.

When trusting the potential factors, there is the danger of betting on the wrong horse. We have to bear in mind that only some of these predictions will be realized, so by selecting the right ones, it is possible to obtain a competitive advantage. The identification of a weak signal is a once-only opportunity. Significant weak signals pass this stage very quickly and change into potential factors and then into dominant views, when more and more people start to recognize them. Most of the existing dominant views have at one time been weak signals themselves. Consequently, the early recognition of the right weak signals provides management with the wild cards of a game—a true strategic advantage with a major impact on business.

References

Ilmola, L., Aaltonen, M., & Autio, E. (2002). Weak signal filters in vision building processes. *Proceedings of the Annual International Conference of the Strategic Management Society.*

Lauraeus, T., & Tinnilä, M. (2003). *Sähköiset palvelut verkostoituneessa monikanavaympäristössä 2020* [working papers] [in Finnish]. Helsinki: Helsinki School of Economics.

Lauraeus, T., & Tinnilä, M. (2004). *Weak signals of e-business in 2020—Heikot signaalit tulevaisuuden sähköisistä palveluista verkostoituneessa monikanavaympäristössä* [research report] [in Finnish]. Electronic Commerce Institute, HSE.

Saarinen, T., & Tinnilä, M. (2002). *Bummat monikanavamalli* [research report] [in Finnish]. Electronic Commerce Institute.

Saarinen T., Kallio J., Tinnilä M., &. Vesa, J. (2002). Customer relationship management in service mediary driven mobile services: Case i-mode. In T. Reponen (Ed.), *Information technology enabled global customer service*. Hershey, PA: Idea Group Publishing.

Tinnilä, M., & Lauraeus, T. (2003). *eBisneksen visio 2020—Tulevaisuuden sähköiset palvelut ja liiketoiminta verkostoituneessa monikanavaympäristössä* [research report] [in Finnish]. Electronic Commerce Institute.

Chapter XXI

Multi-Channel Services of the Future

Jari Manninen, TeliaSonera, Finland

Marja-Liisa Viherä, TeliaSonera, Finland

Abstract

We need to understand people's needs to use technology and what kinds of services, technological infrastructure and legislation are needed to build an open, balanced information society. Future research (i.e., current research aiming at learning about the future) provides the tools for preparing visions and future scenarios and making the future. As an example of making the future, we will present the idea of communication camps, which aim at making multi-channel services more common. In addition, we have written three short stories to illustrate the multi-channel services that arise from people's needs. The stories are based on the use of disciplined technological and sociological imagination (Mills, 1959), and we use them to illustrate the opportunities that multi-channel services offer to everyday operations.

Introduction

The basic infrastructure in Finland is so advanced that it enables multi-channel services. Technology would not stand in our way if we only knew how to use it. The term *multi-channel services* most often seems to be used to refer to content services, such as news delivered and received in various ways. The multi-channel services provided by operators have attracted far less attention. In this article, the term refers to the opportunity of producing, transmitting and receiving various contents through multiple channels.

Demand for a national media culture requires that people know what they need (or want) and also know how to demand it. This, in turn, requires the ability to make one's own productions. It is like the ability to write: only when you learn to write can you understand what you read and demand and appreciate better literature and better texts. Not until multi-channel services become an everyday part of life can we utilize the best, unparalleled opportunity provided by information networks—everyone can be a writer, a cameraperson, and a producer for their own contents. The freedom and joy of speech that does not offend others is everyone's irrevocable right and technology now makes it possible.

Creating a new living and working environment for everyone requires also that we pay attention particularly to the services for small enterprises in order to enable technology to be utilized in full for making everyday life easier. These services cannot be created solely on the terms of the market forces, as it is impossible for small companies to develop the services by themselves. It is also impossible for small, mainly micro companies to do all the background work and to compare and learn to use various new technological applications; that is, to put the technology into full use. They simply do not have enough time. What is also needed is new lines of business and occupational groups (e.g., "IT road service" and network secretaries, who, for instance, attend to the information management, administrative, marketing and information retrieval services of several different micro companies).

We need to understand people's needs to use technology and what kinds of services, technological infrastructure, and legislation are needed to build an open, balanced information society. Future research (i.e., current research aiming at learning about the future) provides the tools for preparing visions and future scenarios and making the future. As an example of making the future, we will present the idea of communication camps, which aim at making multi-channel services more common. We have written the following three small stories to illustrate the multi-channel services that arise from people's own needs. People are the subjects, and the content services have been provided for these subjects. An opposite approach would have resulted in stories about how people use services that are based on corporate and technological interests. Both approaches are needed to identify future opportunities and they should be used in parallel.

Telling stories is an old way of passing on information and understanding. Stories put details together. They help us create images and it is on the basis of images that we make decisions and act (Viherä, 1997). The stories used in business development and strategy work, for example, usually describe the present or the past and are based on accomplished deeds or facts (Aaltonen & Heikkilä, 2003). Scenarios are one way of telling stories about the future. Scenarios often deal with large entities. The following stories are mainly small episodes about the possible future. They are based on the use of disciplined technological and sociological imagination (Mills, 1982), and they aim to illustrate the opportunities that multi-channel services provide for everyday operations.

Story 1: Setting Up a Garden

Pia is sitting on the stairs of her house, thinking about how to put her garden in order for the coming summer. She decides to take action—nothing would happen if she just kept sitting. Pia videotapes the garden and takes ground samples according to a garden adviser's instructions that she found on the Internet. Having completed the groundwork, she sets up a connection to the local garden adviser. She sends the video, information on the ground samples, layout of the garden, and her question, "What should I do?" to the adviser. As it is the gardening season, Pia receives an automatic reply, which states that she will be contacted the next day. However, the reply message contains a Web link that advises Pia to take yet another, more accurate ground sample. The first ground samples gave her information on the quality of soil in different parts of the garden, and now, she can prepare even better for the next day, when she will be contacted by the garden adviser.

Pia talks to the garden adviser first on the phone and tells her about her wishes, while videotaping her garden. The garden adviser, Lea, asks Pia questions and requests her to zoom in to check details, particularly the exact places where she had taken the ground samples the previous day. The weeds growing in these places will give Lea valuable further information. She promises to send a plan to Pia the next day.

On the wide screen of her home, Pia receives an accurate picture of a new garden, information on the plants Lea has chosen, ground improvement plans, and a cost estimate. Pia promises to get back to Lea the next day after having studied the plan in peace and quiet.

Pia supplements the plan with some flowers that are familiar to her from her childhood home. She does not remember their names but manages to find them on the Internet in a catalog of old perennials. She finds them by means of image pattern recognition. As the search criterion, she uses a model she has cut from a digital copy of a childhood black-and-white photo. She also needs the chat service of the 4H Federation to identify one plant.

Lea recommends a good garden worker and gives Pia his contact information. She dictates the invoice to Pia's electronic invoice box (and to her own bookkeeping).

Pia contacts the gardener called Matti and sends him Lea's drawings and plan. Matti is glad to do the job but cannot make it until after a fortnight, which is okay with Pia. Matti promises to order all necessary materials before that and requests that they be delivered in one shipment the day before he begins the work. After a week, a garden center Matti knows well sends a confirmation to his handset that the ordered materials have arrived at the warehouse. The entire delivery has now been confirmed and Matti forwards this information to Pia.

On the agreed date, sand, soil and a large number of different plants are delivered to Pia's garden. At the same time, an invoice for the materials is sent to Pia's electronic invoice box. Pia immediately sets up a connection to her invoice box with her mobile phone and approves the invoices for payment.

Matti arrives in time and does the garden work meticulously and professionally. The coming summer and all summers after that are full of joy and happiness for Pia.

Pia videotapes the progress of the garden work and saves everything in the family's memory box for future generations. She cannot help talking about the garden to her friends. Being also garden enthusiasts, they can enjoy the garden by following its growth over a remote connection.

The following are multi-channel services used in the example:

- transmission of video
- transmission of information on the analysis of ground samples
- streaming camera
- making the plans digitally alive
- image recognition
- multi-channel invoicing, which sends the invoicing information to the invoicer's bookkeeping and the invoices to the invoicee, according to a payment option that suits the invoicee's current situation

- good order logistics systems
- video image saving and distribution service

Story 2: Travel Account

Pia is a university professor of statistics. Pia's university requires that travel accounts be prepared for conference trips to provide information on the conferences to those who have not participated in them. To meet the university's (and her own) requirements and to do the job as well and as easily as possible, Pia has developed her own way of giving travel accounts. When traveling, she carries a handy multi-service device, with which she can videotape, record, type and naturally access the Web.

Even during her trip, Pia is scheduled to teach two groups. One of the groups is dealing with the topics of the conference, so Pia decides to utilize her trip by transmitting lectures to her students as a live transmission. Pia comments on the subject and combines it with her own teaching. She also saves the transmission directly in the network. With the other group, she has taken care of her teaching in advance by means of the Web.

Pia compiles material for her travel account during her entire trip. Her trip thus can be followed in real time, which guarantees a number of recipients for her account. She records the conference lectures and also takes notes during them. She transmits both the recording and her notes immediately to her own protected pages on the Web. If a lecturer has not given his or her permission in advance in the electronic conference program, Pia requests a permission to record the lecture and to publish the recording on the Web.

Pia often also videotapes part of the lecture to make the lecturer more familiar to those who listen to the lecture over a remote connection. She may even interview particularly interesting lecturers, especially if her colleagues in Finland ask her to. Pia receives further questions by text message.

Pia feels that the advanced travel account technology allows her to get much more out of the conference, and it also benefits the entire university. The travel budget of the university is small, and not everyone who is willing can go on a trip. Thanks to the technology, the university maintains its positions as a top university in its field.

The following are multi-channel services used in the example:

- Image
- voice
- video
- text saving and transmission services

Story 3: International School

Pia's son Matias goes to an international school. Each Finnish pupil at the school has his or her own international partner. Matias's partner is Nora from Mexico. Matias and Nora do their homework together with their own team. The team consists of six pupils, three of whom are Finnish. The school believes in the interaction of theory and practice and in the pedagogy of exploratory learning. Teamwork welds the pupils into a community whose members rely on each other. The prerequisite for working is open interaction (Hakkarainen, Lonka & Lipponen, 1999).

Initially, the team members find it difficult to understand each other, since they come from different cultures. Each of them has their own manners and habits. It is perhaps easier for the foreign pupils, as they are in a foreign culture and know to question things, unlike the Finnish pupils. The school has decided to make it systematically easier for everyone to detect differences, to accept them, and to compromise. This is done with the help of multi-channel services. When Nora, for example, notices something she finds as strange, she immediately sends a picture and a question to Matias, who then looks into it. Nora has wondered, for instance, about Finnish people's habits of taking their shoes off when they come to visit, about the Finnish coffee table ceremonies, and so forth. At first, her knowledge of the Finnish language is, moreover, not so good. Nevertheless, Matias finds it easy to understand her question when she sends a picture, taken with a camera phone, of a guest without shoes or of the pile of shoes in the hall, and the question "why." Having received a picture of the service desk of the tax office, Matias understands that Nora is wondering why no one returns her greeting. It is not easy for Matias to answer this, but Nora feels much better when she learns that at least she has not unintentionally hurt anyone. Many similar situations occur, especially during the first months. There often are different image structures behind cultural differences. Language reflects these structures, and even if you have a good command of a foreign language, it is difficult to get behind the image structures (Johansson, 2004). A picture makes this easier.

Teamwork at the school is smooth, as the teams have video cameras, tape recorders, and the like in their use. You often can give a clearer picture of a situation if you videotape it instead of just writing about it. This is self-evident in connection with biology homework, for example. The team keeps in touch by means of group calls, group text messages and group pictures. If, for instance, a team member finds a new plant, he or she sends a picture of it to the other team members. Very often, someone identifies it immediately or can at least quickly edit the question and the related picture for a suitable chat line. It is good to know at once if it is a plant of which one should take a sample.

The team also uses the push talk service, which enables a team member's voice to be heard on everyone else's phone without separate reception. The connective force of the service is huge; the service allows the team members to know almost always where everybody else is and what they are doing.

Matias likes his own school and his own team. In addition to education, the school gives him lifelong friends, and they can always keep in touch, even if they live in different parts of the world.

The following are the multi-channel services used in the example:

- MMS service
- group calls, group text messages, transmission of group pictures
- push talk service

How to Make Multi-Channel Services Part of Everyday Life

The idea of everyone's information society raises the following question: How will people be able to function actively in an information society where they are increasingly bombarded with information?

Learning how to use technology, user-friendly subscriptions, and good online teaching materials do not suffice to answer the question. If learning to use technology is separated from the real situation and need, technology becomes an end in itself instead of a helpful means to accomplish something useful.

Communication Camps

A good way to adopt technology for your own use is to participate in a communication camp. Communication camps have been organized for as long as 19 years. In communication camps, the use of technology is a happy thing that is made part of the everyday activities from the very beginning. What is more important than the use of technology is what you say, how you act and do things together, and how you take care of each other. Communication is almost always an integral part of information and communication technology and it has an important role in all communication camp activities. The aim is not to teach people to master a single means of expression, but to familiarize the campers with as many communication tools and networks formed by them as possible. The key activities in a communication camp include making a video, a radio program, a magazine and food. The campers move in teams from one activity point to another with the aim of getting an overall picture, for example, of the opportunities offered by technology.

You cannot learn to communicate solely by reading books or by learning to know what kinds of communication devices there are. You can only learn to communicate in live situations, when organizing and performing transactions (i.e., through experience). Communication camps serve as examples of situations where a person can learn to communicate both in interactive situations and in action. Very often, when people talk about communication and evaluate communication skills, they forget the ability to make it happen and the ability to use communication in everyday tasks.

In communication camps learning is seen as a human process. In the flexible environment of a communication camp, the campers shape their environment themselves and give it meaning by their actions, manners, and active communication. By observing (even if subconsciously) the impact of their actions and messages, the campers learn new things and meanings. Tacit actions, decision-making procedures, practices and attitudes toward others often have a greater impact on learning than presentations and lectures (Hakkarainen et al., 1999). According to Lonka, lecturing about how to communicate is a waste of time; it is difficult to motivate students, since the subject is abstract and interaction is not only a technological skill, but a part of practical operations. It is these practical operations that constitute a communications camp.

As the camps emphasize democracy and encourage initiative, campers learn to look at things in a new way and to question old practices. They learn to observe their environment and to compare it with the vision of the future—aiming at learning to know about the future.

In communication camps, everyone shares their information and competence with others, as no results would be achieved without cooperation. No one can make it alone. Nevertheless, everyone can operate independently.

The camps offer a person an opportunity to grow toward initiative, sense of community and management of a person's own life. The campers' mutual trust and openness increase in an atmosphere of appreciation. In communication camps, people do not criticize but appreciate each other all the more. Even criticism in the camp media is appreciative criticism.

The main principles of a communication camp arise from social behavior and being together:

- learning by doing
- completing your work
- initiative
- taking care of others
- accepting and bearing responsibility
- everyone is each other's teacher and pupil
- acting together independently
- teamwork
- learning and seeing the overall process
- appreciating others' work

These principles help the campers to become subjects in information society—managers of their own lives together with others.

In the camps, the one who knows more advises the others, or people look for a solution together. This applies particularly to button technology. Those who are more competent show what to do and how to act. Guidance is used to tell especially about matters of which the campers do not have a model in their minds, such as interaction and dealing with everyday logistics. Someone who is more competent than you must be there to tell you that you could have made a call, sent a message or inquiry, and so forth. It is also important to continuously question manners and operations: would it be better in this way or in that way? If a better solution is found, the practice is changed. In addition to teams and leaders, structures and the positions of equipment change. Assuming responsibility teaches you to produce results, even if you do not know how. Experiences of success teach you and give you self-confidence. We must encourage people to try over and over again, praise them for good performances and find great details in weaker performances.

Adoption of Technology

Communal operations refer to the equal opportunity of each member of a community to use, modify and create new information for a jointly maintained operating environment. As a result of new technology, each individual thus has a more active role in the communication of communities. Instead of receiving and waiting passively, they receive information and search for information actively.

In literature on communication science, the theories for studying the use of communication technology distinguish between three approaches to the adoption of technology: rational, social and adaptive (Kallio & Jäkälä, 2000). In the same way that we study the adoption of the actual technology and a service, we also can study how the operations enabled by the service become part of everyday routines.

Rationality in the selection and use of technology mean that we assume communication to be efficient, when the most suitable means for each communication task and situation has been selected by rational criteria. The rational approach also can be called technology-centered, as the use of technology is often explained by technological features. The introduction of technology is based on the technology itself, not on the need for which it would bring new dimensions.

The social approach pays attention to the users of technology. The norms, practices, values, and attitudes prevailing in a social community shape the users' ideas of technology and thereby steer the use and selection of technology in the entire community. Communication camps create new social situations in which the use of technology and the opportunities provided by technology can be studied.

The third adaptive approach studies the use of technology rather as a product of the communal culture than as an individual's choice. The relation between the community and technology is thus always reciprocal, and technology cannot be studied separately from the community. A social community not only defines how technology should be used, but how it also is constantly modified and revamped by the use of technology.

Communication camps allow the multi-channel service to be adopted by the participants so that it becomes part of their everyday lives. The prerequisite for this is that all members of a community participate in the everyday operations in such a manner that the sender and the recipient understand each other and that their services are compatible, as in the case where Pia ordered gardening advice.

Conclusion

From a cultural and communal perspective, a team is the key area for the production and renewal of social information. Communication technology is not only a means for exchanging information and thoughts, but also a phenomenon whose use is determined in social interaction. In communication camps, teams develop and adopt technology, which explains their strong renewing force.

The theory of social influence emphasizes that a social community shapes the ideas of a communication situation and task. Ideas of and attitudes toward a certain technology consist of an individual's experiences of this technology, of his or her usage skills, and of communal factors. An individual's good usage experiences of a certain technology, for example, steers others to use it, too, and vice versa. The progress of multi-channel services depends on positive experiences of use. Saving and distributing self-managed content requires that you can use the service in a familiar environment to perform your own transactions.

Certain kinds of use of technology in a community maintain and strengthen the valid functions and structures of a community but also can dissolve them and replace them with new ones. Technology, as such, does not change the structures of a community, but it can act as a catalyst for the change. In addition to technology, an actual change always requires the input of the social community and communal use of the technology. The use of communication technology, in particular, requires several people with which you can communicate. Otherwise, you will only be a recipient of online information and entertainment.

When developing technological applications, we should consider whether they are intended to be tools for an individual, a certain group or the entire community. It also should be defined whether the aim of a multi-channel service is to enhance communication and individual tasks or to create a sense of community and to change the practices and structures of the entire community. We are certain that a large number of new tasks and products can be found within the area of eBusiness, if we just turn our eyes to people's needs and concentrate on how to facilitate the everyday routines and enhance the quality of life.

References

Aaltonen, M., & Heikkilä, T. (2003). *Tarinoiden voima—Tarinat yritysten elämänlankana.* Helsinki: Talentum.

Hakkarainen, K., Lonka, K., & Lipponen, L. (1999). *Tutkiva oppiminen—Älykkään toiminnan rajat ja niiden ylittäminen.* WSOY.

Johansson, A. (2004). *Kulttuurien taustalla erilaiset mielikuvarakenteet.* Helsingin Sanomat, vieraskynä 30.07.2004.

Kallio, K., & Jäkälä, M. (2000). *Viestintäteknologia työyhteisössä teoksessa*: Valo, Maarit (Ed.): Nykytietoa puheviestinnän opetuksesta, Jyväskylä, Jyväskylän yliopiston viestintätieteiden laitoksen julkaisuja.

Luokola, T. (1989). *Kertomus viestintäleiristä—Tutkimusraportti.* Viestintäkasvatuksen seura ry. Helsinki: Painatuskeskus.

Luokola, T. (1991). *The story of the communications camp.* Helsinki: Hakapaino.

Mills, C.W. (1959) [1976]. *The sociological imagination.* New York: Oxford University Press.

Viestintäleiri – konkreettinen utopia. Viestintäkastatuksen seura ry. Helsinki: Painatuskeskus.

Viherä, M.-L. (1999). *Ihminen tietoyhteiskunnassa. Kansalaisten viestintävalmiudet kansalaisyhteiskunnan mahdollistajan*a (People in information society. The citizens' communication skills and the opening of new prospects for the civil society). Turku: Turun kauppakorkeakoulu.

Viherä, M.-L. (2000). *Digitaalisen arjen viestintä, miksi, millä ja mite*n. Helsinki: Edita.

Chapter XXII

Channel Surfing: The Challenge and Opportunity of Channel Management in a Networked World

Jonathan W. Palmer, College of William & Mary, USA

Abstract

The chapter describes opportunities in channel management at multiple levels, including customer-focused channels providing capabilities in managing customer contacts, sales, and service, as well as procurement and distribution channels. Examples are offered from the U.S. Government's Federal Supply Service, UPS, FedEx, Staples, and PBM Plastics. This chapter summarizes the channels and suggests some likely scenarios for the future, including electronic marketplaces with agents, the evolution of mobile technology, wearable technology, and location-based services. The ability to mix old and new systems and to accommodate legacy as well as recently introduced system components is crucial for success. Perhaps as importantly, the ability to rearrange and reconfigure supply chain channel elements is critical to designing an agile and flexible channel management solution.

Introduction

The typical vision of the TV-watching male with fingers wrapped around a remote control that clicks quietly but continuously in search of the perfect channel may well be a metaphor for those seeking to capture and retain customers in the future.

In the television metaphor, channels represent a series of potential market offerings, distinguished by their content, programming, and advertising. Consumer choices are made as to which channel to use over some time frequency. Channels are easily changed and often are repeatedly changed over the course of a viewing session. Some television hardware facilitates the channel surfing even further with multiple channel presentation on the same screen and the ability to TIVO, or record, the action for later replay.

These customer channels are important, but not the only channels in which today's organizations operate. Firms must identify channels for sourcing people, assets, financing, and all the resources required to successfully compete. The channels of procurement or acquisition also take various forms to facilitate the process of information gathering, negotiation, agreement, settlement, and fulfillment. Channels also can be strong or weak. Within this overall process, multiple channels may be available to support each of the subprocesses (e.g., gathering information from multiple channels about potential employees, products, services, particular firms).

In addition to customer and procurement channels, delivery channels also have increased in number and complexity. Channels for delivery involve a variety of potential fulfillment players, third-party logistics providers, freight forwarders, and customer brokers, all required to move physical product from one location to another. With the advent of digital or information goods, the delivery channels have widened to include electronic transmission of products such as software, news, music, and video. This digital distribution channel also can be extended to provide information support for services, technical support through call centers, and online resources for troubleshooting and information. These delivery channels have another important dimension—the degree of intermediation within the channel. The number of steps and particular companies involved begins to identify the degree of intermediation. At the most basic level, the delivery channel can be intermediated (typically by a broker, dealer, or wholesaler) or disintermediated (with only the principal buyer and seller involved).

This chapter will describe several channel-surfing options facing organizations and provide some insights into what the future may hold. This future is likely one in which the most successful firms will be most adept at managing in a multichannel environment. The chapter is organized first by concentrating on cus-

Table 1. Organizational activities and channel options

Activities	Channel Options	Dimensions	Issues
Customer Contact	In-store, catalog, broadcast, Internet (Web, e-mail)	Attract and Retain Loyalty	Consistency, Pricing
Sales	In-store, catalog, broadcast, Internet (Web, e-mail)	Sales strategy and process	Pricing
Service	Web, phone, in-person	Service Level Agreements, Customer segments	Consistency, Cost
Procurement	Traditional Proprietary (EDI) Open (Internet)	Transaction channel, technology support	Efficiency, price reduction, supply chain performance
Distribution	Intermediated Disintermediated 3/4PL Providers	Structure	Fulfillment performance, Inventory management

tomer-focused channels providing capabilities in managing customer contacts, sales, and service. The following section deals with procurement and distribution channels. The final section summarizes the channels and suggests some likely scenarios for the future and provides insights for successful management in a multi-channel environment.

Customer Contact

Consumers today are faced with a wide array of shopping choices. The information-based society provides methods for purchasing products in a variety of ways—in stores, via catalogs, on television, and through the Internet channel via a variety of equipment, including computers and handheld devices. Retailers also are faced with challenges in diagnosing how and when to use different business models and market mechanisms available to them in an effort to attract consumers and to improve their operations.

There are some fundamental differences between the formats in terms of the degree of interactivity, the level of information supplied about the product, the ability to compare products, and the degree of human intermediation. There is still a significant difference in product description, shopping availability, speed of delivery, and time taken to shop between the four formats (Palmer, 1997, 2000).

The retailing industry is undergoing great change with new retail formats enabled and enhanced by the availability of information and other communication technologies. Catalogs using telemarketing, home shopping cable TV networks, and retailing outlets on the Internet have all expanded the definition of retailing and increased dramatically the diversity of formats available to the consumer.

Channel Differences

The four formats are different in their levels of interactivity, since each is fundamentally based on a different technology. The physical, personal mediation of the in-store format provides the potential for a rich set of product information. The non-store formats have used technology to develop an equivalent to this rich product information. Catalogs provide photographs and textual description and are increasingly moving to CD-ROM and Web sites to enhance product information and the richness of their information. Cable TV provides broadcast sales information but is increasingly making the connection with customers interactive by including consumer phone calls as part of the program. The Web continues the struggle to improve the richness of the information and the user friendliness of the interface. The possibilities are substantial, through product photos, textual description, personal shoppers, and product demonstrations. The underlying technologies (i.e., in-person, telephone, TV, and computer) make significant distinctions in the level of interaction and the product display and definitions. Perhaps these distinctions will be tempered in the future by the similarity in underlying technologies (telephone and computer) used by all of the formats.

The increased availability and use of underlying technologies have created substantial overlap across the formats. The Web has several cyber-equivalents of shopping elements such as electronic shopping carts, video interactive hosts or salespeople, product displays, and video demonstrations. Catalogs have moved to CD-ROM and to the Web; cable TV shopping channels now offer catalogs and Web presence, the Web uses electronic shopping carts and personal shoppers, and in-store formats increasingly include information kiosks and in-store video. This move toward equivalencies of format specific substitutes is a fascinating trend.

Sales and Service Channels

Retailers are faced with an array of possibilities in connecting products with customers. The four retail formats—in-store, catalog, cable television, and the Web—represent different ways of managing the retailer's business and of connecting with customers. Differences among formats make it difficult to find optimal product display and description, product comparisons, delivery timing, sales support, and pricing.

There also appear to be several market mechanisms available through the information technologies that support electronic commerce, including direct search, broker, dealer, and auction (Palmer & Lindemann, 2003). These mechanisms can provide comparative pricing and product information for an individual firm, a group of firms, or across most of the Internet. These technologies support, enhance, and speed up existing commercial transactions (Lee & Clark, 1996) and can influence the cost structure (Lee, 1998). These applications position the World Wide Web (WWW) as a key platform for commercial activity (Alba et al., 1997, Hoffman & Novak, 1996). The electronic markets hypothesis (Malone, Yates & Benjamin, 1987) suggests that the electronic market can be utilized to sell a variety of products efficiently.

The capabilities of electronic markets to search, retrieve, and manipulate information have enhanced specific market mechanisms by incorporating information content on products and vendors as well as historical information on market selling patterns. From a vendor standpoint, the differing market mechanisms now available in electronic markets provide a range of opportunities for selling products. However, some evidence suggests that domination of multichannel retailing format may diminish the pricing efficiency of the Web (Tang & Xing, 2001). The value proposition for buyers and sellers is extended in electronic markets to include elements beyond price, such as trust, speed, liquidity, availability, and quality.

Examining the capabilities of these different market mechanisms is critical to understanding how and when they might best be utilized and which provides the most effective performance on customer value, liquidity, reliability, and accessibility. The capability offered through direct search (e.g., PriceScan) within electronic commerce focuses on consumers seeking standard consumer goods and makes this model very competitive with broker and dealer mechanisms on operational, pricing, and some information efficiencies. Direct search requires consumers to sort through a variety of potential vendors, whereas broker (e.g., Net Markets) and dealer (e.g., Bottom Dollar) models limit the number of vendors and often include suggestions as to preferred vendors.

The auction mechanism offers additional market-making capabilities, bringing buyers and sellers together. Yet the auction mechanism has limitations in the area of order execution, since buyers need to wait for the end of an auction period. This has led firms such as eBay already to move in the direction of providing multiple market mechanisms on its Web site in order to support auctions as well as direct sales of items. Other Web sites, such as Amazon and Yahoo moved the other direction, adding auction mechanisms to their existing fixed-price models.

These four types of electronic market intermediaries are largely reflective of the typology in financial markets. Each type offers different capabilities and all are likely to characterize competitive offerings in electronic markets for the foreseeable future. There is a move toward incorporating multiples of these market mechanisms on the same Web sites.

Case Example: Staples

Staples is a U.S.-based office supply retailer with more than 1,300 retail stores, five e-commerce sites, and a catalog business. The firm has revenues of more than $13 billion U.S. and generates nearly $1 billion U.S. of that through Staples.com. Staples has been successful in leveraging IT for integrating information across sales channels in its own organization, with partners, suppliers, and, most importantly, customers. Utilizing XML and other technologies, Staples has integrated across channels, resulting in improved customer service.

The Staples multi-channel strategy involves supporting all 1,300 retail stores, the catalog, the Internet site, as well as direct sales to large corporate customers. The offerings are broad, with more than 40,000 specialty products, 8,000 core products that are carried in most stores, 100,000 downloadable software titles, and more than 40 business service providers.

Staples identified different needs among its customers and that different customers have different needs at different times. Staples also wanted to insure that customers believed they were interacting with Staples, not the partner providing goods or services. To accomplish this, all channels had to agree upon a common offering. From the technical standpoint, Staples had to get disparate systems to talk, maintain transactional integrity, and develop a system that would support Fault tolerance/7x24 operation. To accomplish this, they developed internal standards, customized Staples.com, rolled out in nearly 1,000 stores by establishing multiple access points in each store (a station where customers could order via the Internet). These access points allowed each store virtually to increase product offerings from 8,000 to more than 50,000. This also allowed Staples to offer business services within a selling context and to offer rich selling content. The access points were used by both associates and customers.

The results of the Staples multi-channel strategy are striking. The typical business customer spends $600 per year at retail. This annual expenditure more than doubles for two-channel shoppers and quadruples for three-channel shoppers. In total, the lifetime value of a three-channel shopper is 4.5X that of a retail-only shopper. By offering three ways to shop and having all channels market the others, Staples has achieved multi-channel success.

Case Example: The UPS and FedEx Battle

Both UPS and FedEx, global package distributors and supply chain and logistics providers, are doing battle for customer channels, especially in the United States. United Parcel Service Inc. acquired the 4,000-outlet Mailboxes Etc. chain in 2001 for $200 million and rebranded most of them as the UPS Store. In addition to the offering of Wi-Fi, copying, and printing for customers, the UPS Stores serve as a new channel for UPS packaging and shipping. FedEx Corp. sees the electronic document delivery services, Wi-Fi Internet access, videoconferencing, copying, and printing capabilities that are offered at 1,200 retail outlets of Kinko's Inc. as a natural complement to its air and ground delivery services.

Case Example: Customer Relationship Management at PBM Plastics

PBM Plastics is a $12 million U.S. manufacturing firm located in Virginia. The firm manufactures plastic-related products for the medical, industrial, and consumer food and beverage industries. Connecting with the retailers that make up the bulk of the firm's customers requires multiple channel capabilities. In addition to advertising in trade journals, attending and presenting at trade shows, and face-to-face sales calls, the firm has implemented a system on the Web site that is designed to assist prospects with designing customized product solutions for their specific applications. The company defines three steps that include concept development, prototype development, and commercialization. In this process, the design properties are chosen entirely by the customer using the self-service Web site, thus consolidating the steps, time, and costs associated with new product design.

Based on the design properties chosen by the prospect, a quotation is automatically generated and e-mailed to the prospect, outlining the costs of engineering/tooling and development time associated with the proposed design. Thus, the prospect receives a confirmation of price within a few days, as opposed to several weeks or months that is typical for most companies developing highly customized products. With the assistance of the Web and its associated systems, PBM Plastics can respond extremely quickly to prospects' requests and offer superior service and responsiveness.

Sample products are necessary and extremely effective for customers to best understand the capabilities of the products and to identify ways the product can meet their distinct requirements. Customers can receive product samples by completing a few simple questions on the Web site and by choosing the particular samples of interest. An online, Web-enabled database has been established to collect basic information about the prospect and the firm that they represent. The information is immediately uploaded to the company's Web site, and it is available to any employee at any time. The information allows PBM Plastics to gain a better understanding of the prospect, their preferences, and various uses of the products, and helps to coordinate the multi-channel sales and service functions.

Requests for samples are typically processed quickly—at least by the next business day. Following the request for samples, prospects receive a series of e-mails requesting a more formal discussion about the products that PBM Plastics offers and how these can assist the prospect. All of these e-mails are automatically generated by the system and also serve as a mechanism for qualifying prospects. The first e-mail thanks the prospect for visiting the Web site and informs the prospect that his or her samples will arrive shortly. Next, a second, automatically generated e-mail is sent to them, asking if they have received the samples that were sent out on a specific date. The e-mail message also explains that the next step in the product development process is to explore ways to adapt the Melt Phase Forming to their application. This involves defining the material, size, shape, and draw (i.e., depth) of the products, and production-quality new products can be developed within two to three weeks.

The progression in customer contact, sales, and service offering on the Web site and support of other channels for customer interaction is a key to the successful multi-channel strategy at PBM Plastics. The firm suggests that its annual expenditures on these systems generate a return of four to five times the investment.

Procurement and Distribution Channels

In addition to coordinating with customers, firms also must look backward in their supply chain to successfully procure items and distribute them both with the company and to end customers. The relationships go beyond the merely dyadic relationships of a single retailer and supplier to the multiple sets of relationships between retailers and their multiple channel partners, including distributors, manufacturers, and, typically, multiple suppliers. Traditional channels to support procurement and distribution include face-to-face interactions and the myriad forms of earlier communication technologies, such as the telephone and fax machine. More recent advances in information technology also have extended the number of available channels.

These IT examples are typically electronic data interchange (EDI) and the Internet, which support electronic supply chains (ESC). These technologies are interorganizational by nature and bring significant potential benefits to the procurement and distribution channels, including improved information exchange (Bakos, 1991; Mukhopadhyay et al., 1995, Raghunathan & Yeh, 2001), purchasing transaction facilitation (Sriram & Banerjee 1994), reduction in inventory (Dresner et al., 2001; Srinivasan et al., 1994), reduction in shipment discrepancies (Srinivasan et al., 1994), and economic benefits (Bergeron & Raymond, 1997; Choudhury, 1997; Clemons & Kleindorfer, 1992; Iacovou & Benbasat, 1995; Lee, Clark & Tam, 1999; Mukhopadhyay et al., 1995; Rao et al., 1995). Issues that have emerged across these channels include volume, profit flow, setup costs, and bargaining power (Barua & Lee, 1997, Johnston & Mak, 2000) and the number of trading partners (Holmes & Srivastava, 1999).

There are various types of basic configurations of ESCs, including electronic dyads, multilateral interorganizational information systems, and electronic monopolies (Choudhury, 1997). ESCs are usually realized in two forms: EDI-based or Internet-based. EDI generally connects firms through proprietary value-added networks (VANs), whereas the Internet generally connects firms through open networks. ESCs have enabled the development of new types of B2B relationships that have the potential to replace or augment traditional intermediaries (Bailey, 1998; Bakos, 1997; Chircu & Kauffman, 1999).

The diffusion level of advanced information technologies, such as EDI and the Internet, in supply chains varies, dependent on many supply-chain-specific characteristics, such as nature of transacted products, level of competition, number of suppliers, intermediaries, and customers. The direct consequence of this uneven diffusion level of ESC is that traditional supply chains (TSCs) and ESCs often co-exist in markets.

An entirely new channel has emerged in the area of procurement—the electronic marketplace of exchange. Organizational options include public exchanges, private exchanges, and consortia-based exchanges. These industry exchanges may allow participants to share industry information, current trends, and best practices. The exchanges support virtual integration with short- and long-term suppliers and enable price, availability, and quality objectives; offer multi-supplier catalog; utilize mechanisms such as auctioning and dynamic pricing; and provide integration with supply chain services such as shipping and fulfillment. The exchanges can identify and profile potential business partners, link up with industry buyers and suppliers, and facilitate the creation of consortia of organizations to leverage aggregate spend. This channel can overcome geographical barriers and provide significant cost savings.

Development of electronic commerce technologies has provided alternative channels for firms to conduct their procurement. However, a recent survey study (Dresner et al., 2003) has found that emerging electronic transaction systems have not replaced traditional transaction systems, although they have been increasingly used. More interestingly, they found that some firms operate both electronic and traditional channels simultaneously. Hence, a fundamental question arises: how are the two channels used differently in transacting different types of products by different types of firms? In many cases, suppliers have to conduct business with buyers over electronic networks or online marketplaces, but they're still stuck using faxes and phone calls to take orders and get them filled.

The advent of electronic commerce channels enabled by these different technologies has prompted many manufacturers to redesign their traditional channel structures by engaging in direct sales. Customer acceptance of these direct channels has been strong in many situations, creating the potential for channel conflict when a manufacturer opens a direct channel to compete with its own retailers (see Figure 1). Many manufacturers who sell directly to consumers online are finding resistance and resentment from the distribution channel partners (Greenberg, 2000). However, in some cases, the mere threat of introducing the direct channel can increase the manufacturer's leverage in the channel (Chiang et al, 2003).

Often, disintermediation does not happen instantaneously, and supply-chain partners find themselves in the uncomfortable position of competing with each other. In some cases, this conflict leads to the dissolution of the relationship. The problem of channel conflict is made more complicated by the introduction of new electronic market intermediaries. The role of exchanges and electronic markets differs in negotiation, deal making, and the role they play in intermediating the channel. The technology-enabled channels have shifted some channel structures and relationships due to the emergence of independent firms utilizing contractual mechanisms to govern both transactions and relationships (Weitz & Jap, 1995).

Case: Federal Supply Service

The Federal Supply Service (FSS) is an intermediary agency between federal government consumers and supplying vendors. The FSS provides supply and procurement services for more than 2,000,000 products. FSS has 312,000

Figure 1. Intermediated (top) and disintermediated supply chains

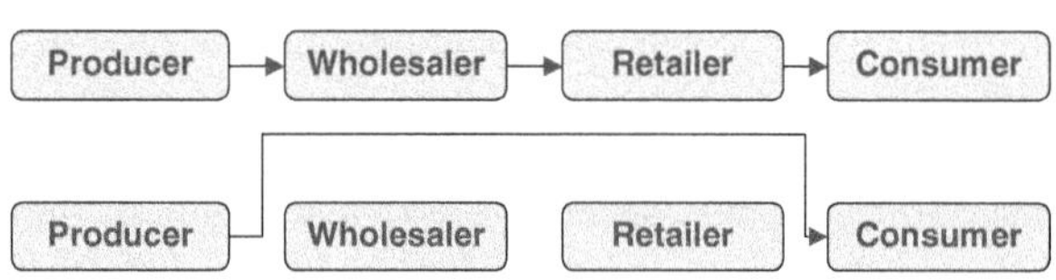

registered users, with the number of browsers using the catalog sites estimated at 1,250,000 annually. The FSS encompass both products procured by traditional systems and those by electronic procurement systems (i.e., EDI and the Internet). Each transaction and order fulfillment record includes selling price, purchase cost, volume, federal purchasing agency, product, transacting market, distribution channel, shipping mode, and fulfillment status. The effect of the electronic channels and disintermediation on prices, price dispersion, and supply chain performance is complex. Use of the Internet to transact purchases appears to improve performance in terms of cycle time and complete shipments, but does not appear to reduce prices. Disintermediation of the supply chain reduces short shipment and lowers prices, but does not improve cycle time (Yao et al., 2001).

Looking to the Future

The final section identifies some possible solutions to the challenges of channel management and offers a metaphor for those attempting to create and manage more successful channels.

Improving Channel Management

As retailers and manufacturers recognize the potential benefits of channel management across the entire supply chain, the focus may shift to better serving the end customer and replacing channel conflict with channel cooperation (Allen, 2001). Organizing communication flows using formal procedures can help reduce instances where one partner may withhold or distort vital information (Mohr & Sohi, 1995). Effective channel management requires clarifying the role of a product within the overall market offering of a firm. This clarification may require a shift in who makes important channel decisions (Cespedes, 1988).

In the future, electronic marketplaces may move to more decentralized market frameworks, envisioning markets with agents. The agent-coordinated functions can allow a procurement agent to access a services directory, intervene in negotiations, and identify potential for virtual aggregation of supply or demand.

The future also will see the evolution of mobile technology. This may take the form of wearable technology that will allow rapid access to customer needs. Voice, data, and video will be supported. Transactions will be without interruption. Location-based services will prevail. Radio frequency identification (RFID) tags will support both improved remote monitoring of tagged items in distribution and procurement channels and allow increased customization of smart paper and packages and increased understanding of consumer decisions. The rapid information transfer made possible by RFID can put channel partners in a position to more rapidly develop products and mass customize them, increasing flexibility in product and service features.

Easy interoperability is an important element to successful supply chains and channel management. The ability to mix old and new systems and to accommodate legacy as well as recently introduced system components is crucial for success. Perhaps as importantly, the ability to rearrange and reconfigure supply chain channel elements is critical to designing an agile and flexible channel management solution.

References

Allen, L. (2000). *Channel conflict crumbles.* Gartner Report.

Bendix, B., Goodman, J., & Nunes, P. (2001). *The channel conflict strategy matrix.* Accenture.

Cespedes, F. (1998). Channel management is general management. *California Management Review, 31*(1), 98-120.

Chiang, W., Chhajed, D., & Hess, J. (2003). Direct marketing, indirect profits: A strategic analysis of dual-channel supply-chain design. *Management Science, 49*(1), 1-20.

Dresner M., Yao Y., & Palmer, J. (2001). Internet technology use across the food industry supply chain. *Transportation Journal, 40*(4), 14-26.

Greenberg, P. (2000, January 4). Manufacturer's beset by e-commerce "channel conflict." *E-Commerce Times.*

Mohr, J., & Sohi, R. (1995). Communication flows in distribution channels: Impact on assessments of communication quality and satisfaction. *Journal of Retailing, 71*(4), 393-416.

The other sort of channel conflict. (2000). *The Economist, 355*(8171), 4.

Palmer, J. (1997a). Electronic commerce in retailing: Differences across retailing formats. *The Information Society, 13*(1), 97-108.

Palmer, J. (1997b). Retailing on the WWW: The use of electronic product catalogs. *International Journal of Electronic Markets, 7*(3), 6-9.

Palmer, J. (2000). Electronic commerce in retailing: Convenience, search costs, delivery and price across retail formats. *Journal of Information Technology & Management, 1*(1), 25-43.

Palmer, J., & Lindemann, M. (2003). Business models and market mechanisms: Evaluating efficiencies in consumer electronic markets. *DATA BASE, 32*(4), 23-38.

Palmer, J., et al. (2000). Online grocery shopping around the world: Examples of key business models. *Communications of AIS, 4*(3).

Tang, F., & Xing, X. (2001). Will the growth of multi-channel retailing diminish the pricing efficiency of the Web? *Journal of Retailing, 7*(3), 319-330.

Weitz, B., & Jap, S. (1995). Relationship marketing and distribution channels. *Journal of the Academy of Marketing Science, 23*(4), 305-320.

Yao, Y., Palmer, J., & Dresner, M. (2002). Impacts of electronic commerce on supply chain performance, structure, and pricing. *Proceedings of the ICIS 2002,* Barcelona.

About the Authors

Timo Saarinen, PhD, is a professor of information systems and electronic business at the Helsinki School of Economics, Finland. He is also Vice-Rector of Research, Chairman of the Department of Management, Director of the Doctoral Program, and Chairman of the Electronic Commerce Institute. He earned his PhD in 1993 and has published in major journals such as *MIS Quarterly*, *Journal of Management Information Systems*, *Information & Management*, and *Journal of Strategic Information Systems*. His research interests include economics of information systems and the development of efficient market-driven business applications, with a recent focus on the multi-channel environment of electronic commerce.

Markku Tinnilä, PhD, is director at LTT Research, Ltd. of the Helsinki School of Economics (HSE), Finland. He earned his PhD in the field of business and service processes (1997) at HSE. He has published in the *Business Process Re-engineering and Management Journal*, *International Journal of Service Industry Management*, *Journal of Strategic Information Systems*, *International Journal of Logistics Management*, and *Information Technology and Tourism*, in addition to numerous conference papers and many monographs. He is responsible for the development of research projects at the LTT and has acted as project leader in several multi-year, multiple-customer projects in different fields of electronic commerce and other businesses. His research interests include efficient business and service processes, multiple access technologies in e-commerce, and business models. He is an Adjunct Professor at HSE in the fields of services of electronic business and business process management.

Anne Tseng is a project manager at LTT Research, Ltd. of the Helsinki School of Economics, Finland. She has authored and co-authored several reports in finance and electronic commerce for LTT Research including a forthcoming article in the *Business Process Management Journal*. Ms. Tseng received her MBA from the MIT Sloan School of Management and an AB in economics from the University of Chicago.

****

Tomi Dahlberg, PhD, is a professor of information systems science at Helsinki School of Economics, Finland, and Chairman of the Board at three Finnish companies. He has worked for more than 20 years in various senior executive positions in software industry, banking, and telecommunication, and nearly 10 years in academia. His academic work focuses on mobile and electronic commerce services, IT governance, strategic management of electronic and mobile commerce, and on IT-related innovation management and new business development. Dr. Dahlberg has an MBA (1980) and a PhD (1991, Econ.) from Helsinki School of Economics, the latter in information systems science. His PhD dissertation, *The Effectiveness of Report Format and Aggregation*, was awarded the prize of the best IS/IT PhD dissertation in Finland (1991).

Kari Elkelä, Master of Social Sciences, is a development manager at Finland Post Corporation, where he has carried out empirical research for more than 20 years. He is also a post-graduate student at the Helsinki School of Economics, Finland, and has specialized in reception channel preferences in consumer marketing.

Veikko Hara is Senior Vice President and Chief Technical Officer (CTO) at TeliaSonera, Sweden, and is responsible for Corporate Strategic leadership of long-term service development.

Jukka Heikkilä (aka Jups) holds a PhD (Econ.) in information systems from the Helsinki School of Economics, Finland. He is a professor of electronic commerce at the University of Jyväskylä, Docent of innovative technologies at Helsinki School of Economics and Business Administration, and Docent of information systems at the University of Turku. His current research focuses on the problem of adopting and implementing innovative technologies to support businesses and their consequences on organizations. A topical theme is to find out, how to design and handle distributed activities over a network of companies. He has run numerous research and consultancy projects in cooperation with other universities and industries.

Marikka Heikkilä is a lecturer in the Department of Computer Science and Information Systems, University of Jyväskylä, Finland. She received an LicSc (Econ.) degree in Information Systems from Helsinki School of Economics (1996). She has worked as an assistant, lecturer, and researcher at the Helsinki School of Economics and the University of Jyväskylä since 1988. Her recent articles have been published in several book and conferences such as IRIS and IFIP I3E. She has also participated in many national applied research projects. Her areas of research interests are ICT supported cooperation, business networks, and coordination.

Magnus Holmqvist, MSc, is an Industrial PhD Candidate at the Göteborg University and currently works with business innovation at Volvo Information Technology, Sweden. He has extensive experience from global supply chain management with after-market business as a specialty. Holmqvist's research interest concerns integration of complex systems, and he is an appreciated guest lecturer also in subjects such as process development and project management.

Jukka Kallio is vice president of LTT Research, Ltd., an affiliate of the Helsinki School of Economics, Finland. He has published in many major conference proceedings and journals, including *Electronic Markets, Information, Technology and People* and the *Journal of Strategic Information Systems.*

Stefan Klein is the John Sharkey Professor of electronic commerce in the MIS Department, University College, Dublin, Ireland, on leave from his position as Director of the Department of IS at the University of Muenster, Germany. He has held teaching or research positions at the Universities of Linz, Austria, Koblenz-Landau, Germany, and St. Gallen, Switzerland, at Harvard University, the German National Research Center for Computer Science (GMD), and the University of Cologne. His research interests include the impact of ICTs on industry structures and interorganizational arrangements, electronic business strategies, and information management.

Tuomas Kokkonen, MSc (Econ), is a logistics development manager at Finland Post Corporation, Finland. He is also a post-graduate Student at the Helsinki School of Economics and has specialized in research on distance sales and logistics services.

Juha Laine, JD, is a professor at the Helsinki University of Technology, Finland, and focuses on legislation and regulation as it pertains to e-business. He has published in journals such as Finland's *Defensor Legis*. He received his doctorate in law from the University of Helsinki.

Iris Theresa Lauraeus-Niinivaara is a PhD student finishing her doctoral dissertation on *Consumer Buying Behaviour and Information Search in the Multi-Channel Environment* in information system science at the Helsinki School of Economics, Finland. Her research interests include consumer buying behavior, information search processes, and uncertainties of search and buying in multi-channel environment, networked business models, and weak signals, among others. She is a member of the Graduate School for Electronic Business and Software Industry and a member of the Multi-Disciplinary Research Group on Electronic Commerce and Services. Before her doctoral studies, she worked as a consultant and director of a Finnish Internet office.

Mirella Lähteenmäki is a doctoral candidate and coordinator at the Center of Direct Marketing Excellence, associated with the Department of Marketing at Helsinki School of Economics, Helsinki, Finland.

Niina Mallat is a researcher and PhD candidate in the Department of Information Systems Science at the Helsinki School of Economics, Finland. Her research interests are focusing on the topics of electronic and mobile commerce, electronic payment systems, and technology adoption.

Jari Manninen, MSc (Tech), is development director of the Information Society Unit of TeliaSonera, Finland.

Marko Merisavo is a researcher and project manager at the Helsinki School of Economics, LTT Research, Ltd. in Helsinki, Finland. His work focuses on the use of e-mail, the Internet, mobile media, and digital TV in customer relationship marketing. He is also developing marketing metrics to track the effectiveness of multi-channel campaigns and CRM efforts. His articles have been published in the *Journal of Advertising Research* and *Journal of Product and Brand Management.*

Karri Mikkonen is director of corporate strategy at TeliaSonera, Sweden. He has been involved in various corporate strategy-related tasks since the company was formed as a result of a merger between Finnish and Swedish telecom operators. His strategic interest in the telecom industry is based on its challenging aspects of change. Since 2000, he continues PhD studies in the strategy area with special interest in telecom competition dynamics and timing through business model theories. Karri has studied Dutch and European integration at the University Amsterdam (1993), cognitive artificial intelligence and philosophy at Utrecht University (1993-1995), and industrial management and electronics manufacturing at Tampere University of Technology (1995-2000). Before that, he spent his time in the mountains of Corsica, the forests and deserts of Africa, and the jungles of the Amazon with parachute and jungle regiments of the French Foreign Legion (1988-1992).

Claas Müller-Lankenau is a researcher and PhD Student in the Department of Information Systems, University of Muenster, Germany. He studied information systems at the University of Muenster and Management Science at Warwick Business School. As holder of a Heinz Nixdorf Scholarship, he subsequently spent one year in Asia and worked at Merck KGaA Vietnam, where he managed an information systems development project. His research interests include interorganizational information systems and networks, e-business and multi channel strategies, and electronic customer interactions. He has been engaged in research and consulting projects on electronic procurement, interfirm networks, multi-channel management, and Web evaluation.

Heikki Nikali, PhD, is research director at Finland Post Corporation, Finland. His area of interest includes factors that direct the communication behavior of organizations and individuals, particularly, competition between the print and electronic media. He has published many international articles.

Andreas Nilsson is a PhD candidate at the Department of Informatics, Göteborg University, Sweden. He lectures in the design of mobile information technology. His research interests focus on the interplay between spectators and large events, the ways that spectators interact with information and other people in situ, and how technology can be designed to support spectator practices.

Janne Orava, MSc (Econ.), is a researcher at LTT Research, Ltd., part of the Helsinki School of Economics, Finland. Previously, he worked as a product manager and Sales Manager at Smartner Information Systems Ltd. (www.smartner.com), and as a Consultant at Adcore Finland. He has done several research and market analysis reports and business projects on topics related to digital TV, digital communities, mobile and Internet services, and portals.

Jonathan W. Palmer is an associate professor of operations and information Technology at the College of William & Mary, USA. His current research interests include supply chains and electronic commerce. He has published more than 30 articles appearing in leading journals such as *Information Systems Research*, *Communications of the ACM*, *International Journal of Electronic Commerce*, *European Management Journal*, and *The Information Society*, and currently serves on the editorial board of *International Journal of Electronic Markets* and *European Journal of Information Systems*. He has consulted with a variety of organizations, including Honeywell, Nokia, Lockheed Martin, Retex, Food Logistics, and the U.S. General Services Administration.

Ken Peffers, PhD (Purdue, 1991), is an associate professor of MIS at the University of Nevada, Las Vegas, USA. His current research focuses on making the right investments to build information systems for the firm, developing system features for maximum benefit to the firm, and managing the portfolio of IS ideas and projects. His articles have appeared in many of the world's leading information systems journals. He is editor-in-chief of the IS journal, *JITTA*, accessible at http://www.jitta.org/, and is a member of the Sault Ste. Marie Tribe of Chippewa Indians.

Tommi Pelkonen, MSc (Econ.), is a Doctorate Student at the Helsinki School of Economics, Finland (www.hkkk.fi). In his professional career, he has worked as management consultant, specializing in mobile telecommunications, internationalization, and business strategy formulation at Satama Interactive (www.satama.com), a European digital services firm. He has worked on multiple mobility-related business projects. Prior to Satama, Pelkonen worked as project manager and senior researcher in LTT Research, Ltd. (www.ltt-tutkimus.fi), analyzing the developments in the Finnish interactive service provision markets. This topic also forms the theme of his doctoral dissertation. Pelkonen has authored several publications of the Finnish digital media landscape. His latest report (2003) was analyzing the Finnish digital media markets. In addition, Pelkonen has worked as IT-project supervisor and lecturer in the Information Technology Program (ITP) at the Helsinki School of Economics (itp.hkkk.fi).

Esko Penttinen is a soctoral student in Information systems science at the Helsinki School of Economics, Finland. He holds an MSc in Economics from the Helsinki School of Economics and a diploma from the Dijon Business School, France. His research interests include the bundling of products and services, and the transition from product-oriented thinking patterns toward services-based concepts, especially in the traditional manufacturing industry. Currently, he is conducting research for his doctoral dissertation on the fundamental organizational changes in business relations and product offerings initiated by new technologies.

Mika Perttula, BSc (Econ.), is a consultant in his own consultancy company (Unique Element Ltd.) and a Researcher at LTT Research, Ltd. (www.ltt-tutkimus.fi). Previously, he has worked as a director of professional services at Smartner Information Systems Ltd. (www.smartner.com) and as a project manager at Adcore Finland. He has done several research, market analysis, and business development projects for Finnish companies on topics related to digital TV, software business, mobile, and Internet services.

Kalevi Pessi, PhD, leads the Business Technology Research Group at the Viktoria Institute, Sweden. He is head of the IT management master program at the IT-University in Göteborg. Pessi did his PhD thesis on IT-management, and current research interests concern business value and architectural design. He has been managing director of the Viktoria Institute and knowledge management director at Guide, a large Swedish consultancy firm, and has wide experiences from trade and industry.

Anu Raijas, PhD (Consumer Econ.), has been a University Lecturer in consumer economics at the University of Helsinki, Finland, Department of Economics and Management, since 2002. Her primary research fields are consumption, shopping behavior, electronic shopping, and use of information and communication technology. Her articles have been published in the *International Journal of Consumer Studies*, *Journal of Product & Brand Management*, *Journal of Retailing and Consumer Services*, *The European Retail Digest*, and *The International Review of Retail, Distribution and Consumer Research*.

Risto Rajala, MSc (Econ.), is a researcher and PhD student at the Helsinki School of Economics, Finland, majoring in information systems science. His current research is focusing on business models, revenue logics, and networked product development strategies and methods of software companies. Prior to his doctoral dissertation project, he worked as a software development practitioner at the Nokia Research Centre and Fujitsu ICL. During his PhD studies, he has published academic articles and conference papers on software business models and revenue logics of software companies and the formation of software product and service offerings. His latest research activities include participation in a research project called "SME Innovation and Growth in Network Economy." The project examines how science- and technology-driven SMEs can increase their probabilities of successful innovation commercialization and rapid market growth through strategic networking.

Matti Rossi, PhD, is acting professor of information systems and director of the Electronic Business program for Professionals at the Helsinki School of Economics, Finland. He has worked as Research Fellow at Erasmus University Rotterdam and as visiting assistant professor at Georgia State University, Atlanta. He received his PhD in business administration from the University of Jyväskylä (1998). He has been the principal investigator in several major research projects funded by the technological development center of Finland and Academy of Finland. His research papers have appeared in *Journal of AIS*, *Information and Organization*, *Information and Management*, and *Information Systems*, and at more than 20 conferences such as ICIS, HICSS and CAiSE.

Virpi Kristiina Tuunainen, PhD, is a professor of information systems science and director of GEBSI (Graduate School for Electronic Business and Software Industry) of the Helsinki School of Economics (HSE), Finland. She received her PhD from HSE. She has been a Visiting Scholar at the Erasmus University Rotterdam (The Netherlands), University of Texas at Austin (USA), and a Visiting Researcher at the Business School and at the Department of Computer Science of University of Hong Kong. Her research focuses on electronic and mobile business, software industry, and economics of IS. She has published in journals such as *MIS Quarterly*, *Communications of the ACM*, *Journal of Management Information Systems*, *Journal of Strategic Information Systems*, *Information & Management*, and *Information Society*. Her research papers also have appeared in a number of international conferences. She is a member of the editorial boards of *Information & Management*, *e-Service Journal*, *JITTA*, and *EM - Electronic Markets*.

Tuure Tuunanen, MSc (Econ.) (Helsinki, 2001), is finalizing his PhD in information systems science at the Helsinki School of Economics, Finland. His doctoral dissertation focuses on requirements elicitation methods to support idea generation and software development in wide audience end-user environments, such as JAVA applications for mobile phones and digital TV terminals. He is currently a Research Fellow at the Graduate School of Electronic Business and Software Industry. His research has appeared in *Journal of Management Information Systems*, *Information & Management*, and *Journal of Information Technology Theory and Application*.

Eric van Heck, PhD, is a professor of business administration at Erasmus University's Rotterdam School of Management, The Netherlands, where he teaches in the international MBA program and in the Global eManagement program. His research concentrates on the design of electronic markets and on the design of business modularization. In his research, he helps companies to develop innovative, electronic auctions. van Heck has published in journals such as the *California Management Review*, *Communications of the ACM*, *Decision Support Systems*, *European Management Journal*, *Harvard Business Review*, *Information Systems Research*, and *WirtschaftsInformatik*. He also has written or edited 12 books, among them *Making Markets* (Harvard Business School Press, 2002).

Jarkko Vesa is a Research Fellow at Helsinki School of Economics, Finland, in the Information Systems Science Department. During the past five years, he also has worked as a freelance journalist for various business and IT magazines and newspapers in Finland and works frequently as a management consultant for leading Finnish companies. Jarkko's research concentrates on multi-channel services and technologies, focusing on the transformation of mobile services industry, where he has analyzed the differences in the structure and success of mobile service in Japan and Europe. His book on mobile services in a networked economy is available from Idea Group Inc.

Marja-Liisa Viherä, PhD, is research director of the Information Society Unit of TeliaSonera, Finland.

Janne Vihinen, MSc (Tech.), is a PhD Student at the Helsinki School of Economics (HSE), Finland, majoring in information systems science. His current research focuses on mobile services and mobile games and mobile operator's business models. Prior to his doctoral dissertation project, he was a sales manager at the Fujitsu ICL and partner manager at Sonera.

Kai Wehmeyer is a researcher and PhD Student at the Department of Information Systems, University of Muenster, Germany. He studied marketing and IS at the University of Muenster and currently industrial and organizational psychology at the University of Hagen. He is involved in research projects on interactive and direct marketing, electronic business, information management, and multi-channel strategies. His research interests include electronic business, mobile commerce, interactive marketing, trust, and consumer/IO psychology.

Anssi Öörni, PhD, is assistant professor of information systems at the Helsinki School of Economics, Finland. He received his MSc and PhD in economics at the Helsinki School of Economics. He has published in the *European Journal of Information Systems*, *Information Technology and Tourism*, and the *Journal of Travel Marketing*. His current research interests are in consumer behavior in electronic markets and diffusion of mobile payment systems.

Index

www.ingramcontent.com/pod-product-compliance
Ingram Content Group UK Ltd.
Pitfield, Milton Keynes, MK11 3LW, UK
UKHW050947290926
15470UKWH00005B/1698